海外特色餐厅
UNIQUE DESIGNED RESTAURANT

深圳市海阅通文化传播有限公司 编

凤凰出版传媒集团 | 凤凰空间
江苏人民出版社 | IFENGSPACE

CONTENTS

CONTENTS

迪拜 Nobu 餐厅

Project > 项目名称 > 迪拜 Nobu 餐厅

Architect > 设计公司 > Rockwell Group

Location > 项目地点 > 迪拜

Built area > 项目面积 > 1069.5 m^2

Photographer > 摄影师 > Eric Laignel

Creneau International
ideas of the anti-ordinary

项目介绍 PROJECT DESCRIPTION

In Dubai, Rockwell Group has created an extension of the Nobu brand for the first Middle East location. The design is an evolution of many of the concepts developed for the flagship Nobu 57, such as the emphasis on craftsmanship, natural materials and storytelling. Large-scale computer generated woven panels surround the restaurant walls and ceiling, creating a fluid, curvilinear environment for dining.

罗克威尔设计公司在迪拜为 Nobu 打造了第一个位于中东地区的餐厅。这个项目的设计传承了 Nobu57 旗舰店的许多设计理念，例如重视手工艺、天然材料的运用以及设计表达。餐厅墙上和天花板上大面积的编织面板围绕着餐厅，营造了一个流畅、富于变化性的用餐环境。

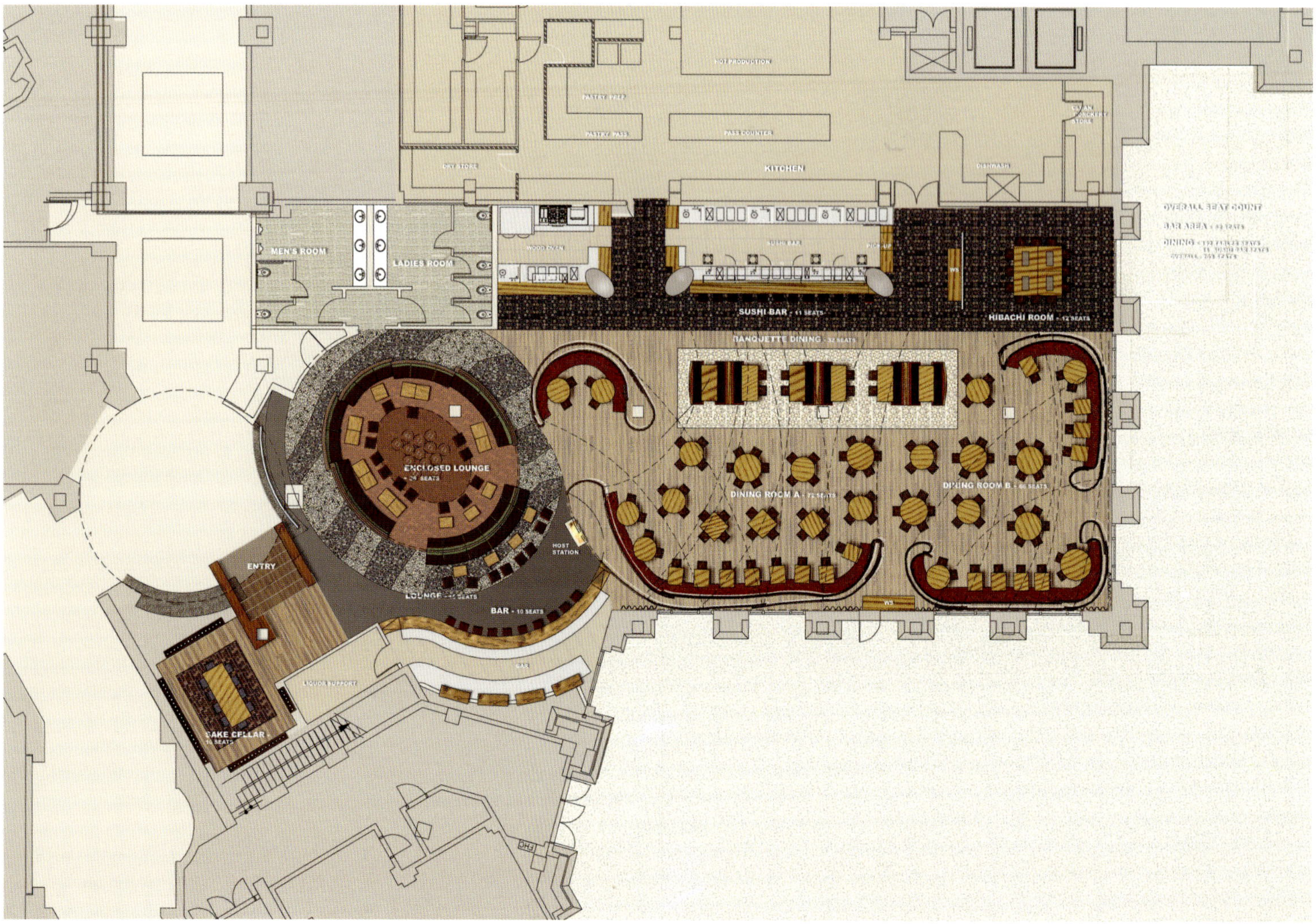

Three dimensional woven abaca panels surround the main dining room, creating an experience akin to being immersed under an ocean wave. Guests sit on uplift woven banquettes formed by these colorful panels, cascading down from the ceiling towards the floor.

立体感强的麻蕉编织板环绕主餐厅，营造了一种身处海浪中的氛围。客人们坐在编织板构造的走廊中，放眼望去，从天花板到地板满眼都是绚烂的色彩。

A 60-foot long curved river rock wall and laser cut metal screen of cherry blossoms mark the entry to the bar. The bar itself is made from 3 inch thick solid timber planks with a backlit onyx front. For the lounge, Rockwell Group has developed curved translucent flower panels with cast images of custom large scale cherry blossom photographs by Michael Palladino, which form an elliptical shape around the center of the space to form an area for private events.

一面 18 米长、弯曲的石墙和一块激光切割金属做成的美丽的樱花屏风向人们展示酒吧的入口。吧台用 8 厘米厚的木板做成，表面镶嵌带背景光的缟玛瑙。休息室设计了一些半透明的弧形花纹装饰板，上面贴有定制的由 Michael Palladino 拍摄的大幅樱花照片，整体呈椭圆形，环绕整个空间，构成私人会谈区域。

Juxtaposed with the curving fluidity of the abaca panels and their translucent forms is the rectilinear sushi bar. All surfaces of the sushi bar are encased in black bamboo embedded terrazzo. A white bamboo proscenium forms a theatrical backdrop for the sushi chefs at work. This space is flanked on one side by a frosted blue mirror service bar, and on the other by a curved plaster wall encrusted with metalwork flowers encasing the private dining room.

与形状起伏且半透明的麻蕉板平行的是直线型的寿司吧台。吧台的表面被紫竹嵌入式水磨石覆盖，白竹的背景为寿司厨师创造了一个恰似舞台的工作环境。包间两侧一边是蓝色磨砂玻璃做成的吧台，另一边是呈弧形的灰泥墙，手工制作的金属花缀满墙面。

Wooden arches greet the guests entering the private dining room. The walls of this space are covered with glowing sake bottles and custom artwork on two sides. Overhead the metallic ceiling in punctuated with a glowing composition of custom suspended pendants.

穿过木质拱门就到达了私人用餐室。这里的墙壁挂满了发光的酒瓶和定制的艺术品。金属天花板中穿插了很多发光的吊饰。

“纸鱼” 海鲜餐厅

Project > 项目名称 > “纸鱼”海鲜餐厅

Architect > 设计公司 > Clive Wilkinson 设计公司

Location > 项目地点 > 美国加州

Built area > 项目面积 > 718 m^2

项目介绍 PROJECT DESCRIPTION

CWA worked with the client to brainstorm a vision for the restaurant and proposed a total design concept which included all visual elements. Designers proposed the name “Paperfish” to underscore the restaurant’s function and to align with the concept of treating the plain white architecture as thin leaves of curving drywall to evoke fluidity, water flowing and pooling, and to give the sense of being immersed below the sea. The word “paper” also evoked fish cooked “en papilotte”, and this has become the restaurant’s signature dish.

Clive Wilkinson 设计公司与客户一起构思设计方案，考虑了所有的视觉元素。他们将餐厅的名字命名为“纸鱼”，强调了餐厅的功能，同时暗含建筑特点 —— 简单的白色风格，薄叶片一样的弧形墙像水一样“流动”，使人仿佛置身于海底。“纸”这个词能引起人的遐想，用纸包裹着做鱼，这也成为了这家餐厅的招牌菜。

Paperfish Restaurant

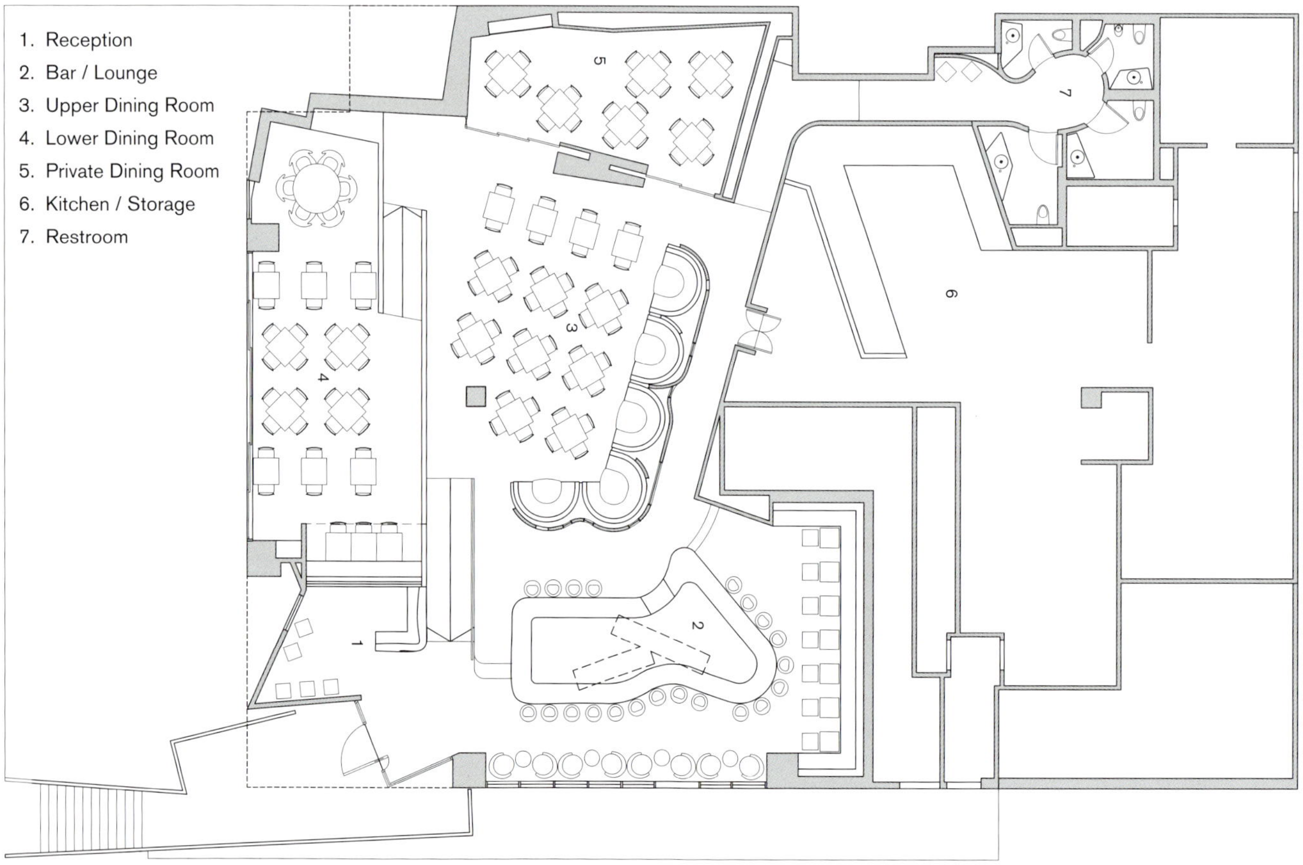

The complete brand identity was developed simultaneously with the architectural design. Designers researched plates and uniforms, glassware and upholstery, door handles and light fixtures, logos and fonts all at the same time. The restaurant experience needed to be effortless and smoothly integrated, with an architecture that could form a subtle and calming backdrop to the dining event. The environment evokes the ocean in movement, but the effect is almost subliminal as the color blue is absent. The shades of red to orange relate to color changes on the body of a carp fish.

完整的品牌形象与建筑设计同样重要。设计师在设计中同时考虑了盘子、制服、玻璃制品、软装、门把手、灯具和标志。在餐厅就餐应该放松，很好地融入环境，因此设计师打造了一个精致、舒适的就餐环境。这里能让人在走动时联想到海洋，但这是在人的潜意识里，而并没有使用蓝色来渲染。红色和橙色的色调选用源自来自于鲤鱼身上颜色的变化。

The experience begins at the reconfigured entrance off Maple Drive. The landscaped verges of the stone clad building separate you from the quiet neighborhood street, and you are drawn up the stairs into a world removed from the hectic city of Beverly Hills. On entering, the restaurant unfolds as a series of separate but connected spaces. The paper bar is an irregular swirling form that harbors nests of barstools and lounge seating, with a quiet rear alcove for enjoying cocktails and occasional dining.

Hanging over the bar like driftwood, a large custom designed functional chandelier of glass and wine storage is illuminated from behind to dramatic effect. Materials are natural feeling: a maple hardwood floor flows into the bar walls, shades of metal gray and orange-flecked chairs echo fish colors. The classic hanging Danese light fixtures evoke lobster pots. Overhead, the ceiling begins a play of fluid shapes that evoke fish movement, water and torn paper, all made from the simple palette of white painted gypsum board.

Minimal pendant spotlights illuminate the cloth covered tables which are surrounded by organic shaped chairs in grey fabric alternating with red. Concealed from view, strip lights within the ceiling subtly change the space as the evening begins and underscore the theme of fluidity and other world.

枫林道上的入口处重新被设计，石头饰面建筑的景观让人置身于宁静的街道，跨上楼梯，从比佛利山庄喧嚣的环境中进入另外一个世界。餐厅被分隔成一系列的小空间，但又各自相连。纸鱼吧的形状是不规则的漩涡状，摆放着高脚凳和座椅，后面比较安静的凹形房间用于鸡尾酒会和特殊场合的晚宴。

一盏定制的玻璃大吊灯和酒柜的背光效果像舞台一样充满梦幻，像浮木悬在吧台上方。天然材料的运用提高了舒适感，地板和吧台的墙壁都用枫木铺装，灰底橙点的金属椅子和鱼的颜色相呼应。古典的 Danese 灯具像一个个捕龙虾的笼子。流线型的天花板使用简单的白色石膏板，象征了鱼的流动性，同时使人联想到水和撕碎的纸片。

最小型号的吊灯照亮了布面桌子，它周围有许多红灰相间面料的椅子。日光灯隐藏在天花板中，巧妙地为这里营造了傍晚的气氛，强调了空间的流线感和立体感。

EXIT

+green 餐厅

Project > 项目名称 > +green 餐厅

Architect > 设计公司 > Sinato 建筑事务所 （Chikara Ohno）

Location > 项目地点 > 日本

Built area > 项目面积 > 111.53 m^2

Photographer > 摄影师 > Toshiyuki Yano

项目介绍 PROJECT DESCRIPTION

An organic restaurant located on Jiyu Street, which is a short walk from Komazawa Park, one of the biggest parks in Tokyo. The restaurant is on the ground floor (half-underground) of a three-story apartment building (reinforced concrete). Since it is located in a residential area, there are not so many people on the street. But on the weekend, there are many people walking who go to Komazawa Park for a picnic, especially in good seasons.

+green 餐厅位于东京鲫鱼大街，从东京最大的驹泽公园步行到这里只需要很短的路程。餐厅位于一座三层公寓的底层（地下半层）。因为这里是居民区，街道上没有太多行人。但在周末尤其是天气好的时候，很多人会去驹泽公园野餐。

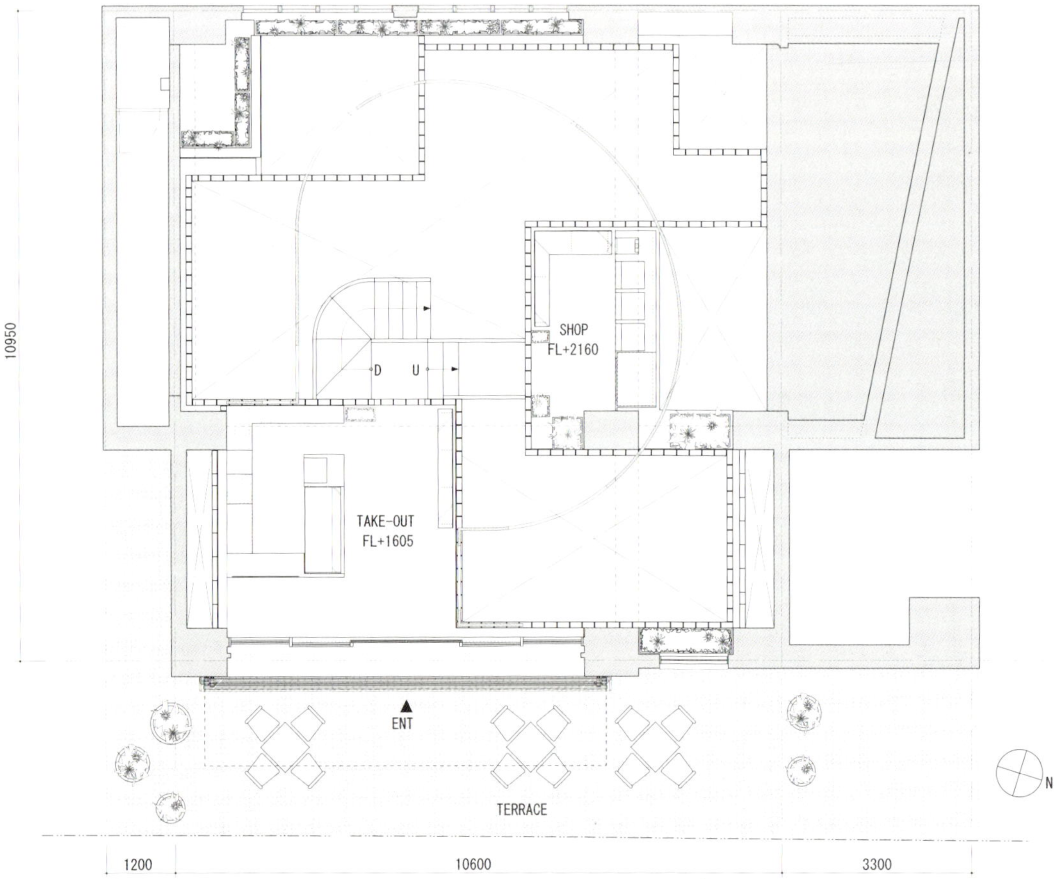

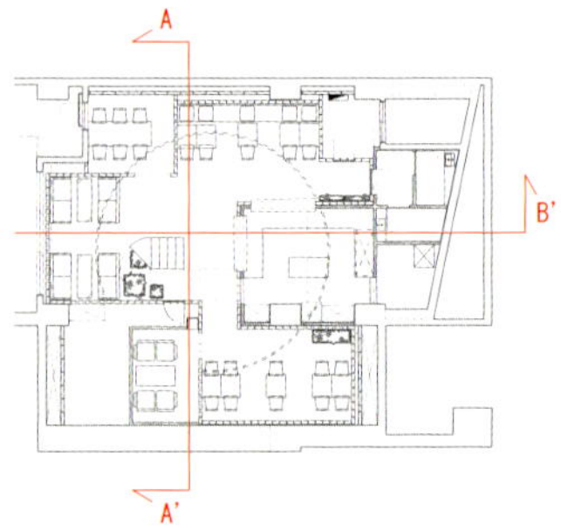

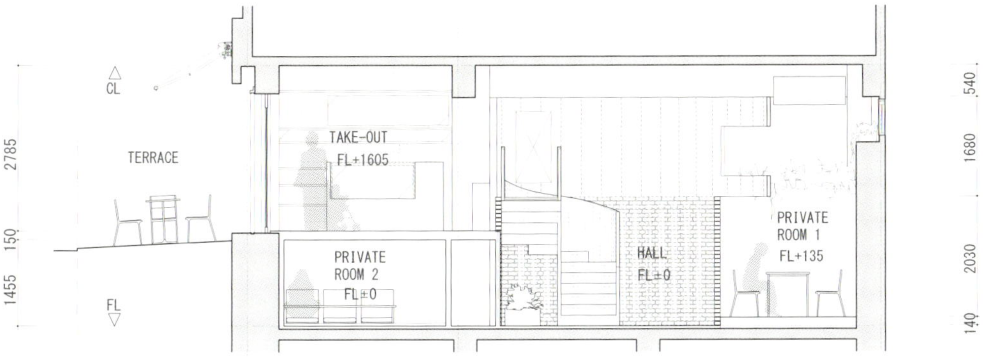

Designer arranged the restaurant on the half underground floor, the takeout place in front of the entrance at same level as the ground and the shop place diagonally away from the takeout place not to hide the restaurant from the entrance and to get sunlight on the restaurant floor from facade opening. The shop place is 0.56m above the takeout place in order to arrange the kitchen under it. There are 3 floor levels for each function, so people move up and down in this space.

设计师将餐厅安排在地下半层，外卖区位于同一层的入口前方，商店区在外卖区的斜对面。没有遮挡入口处的餐厅区，有利于这个区域的采光。商店区比外卖区高 0.56 米，便于下方厨房的设置。每个功能区都有三层楼，客人们从这里上上下下。

It has three functions such as a takeout, an organic food shop and a restaurant.The most distinctive feature of the premises lies in its floor level, which is 1.61m below the ground level of entrance. And the height of the premises is 4.39m.To study the arrangement of three functions in such a unique space was the starting point of our design.

餐厅有三个功能：外卖、有机食品店和餐厅。设计的最大特点是：一层入口的下方有 1.61 米高的空间，房子总高度是 4.39 米。在这样一个富有特色的空间中，设计的出发点是对餐厅三个功能的实现。

You
are
what
you
eat

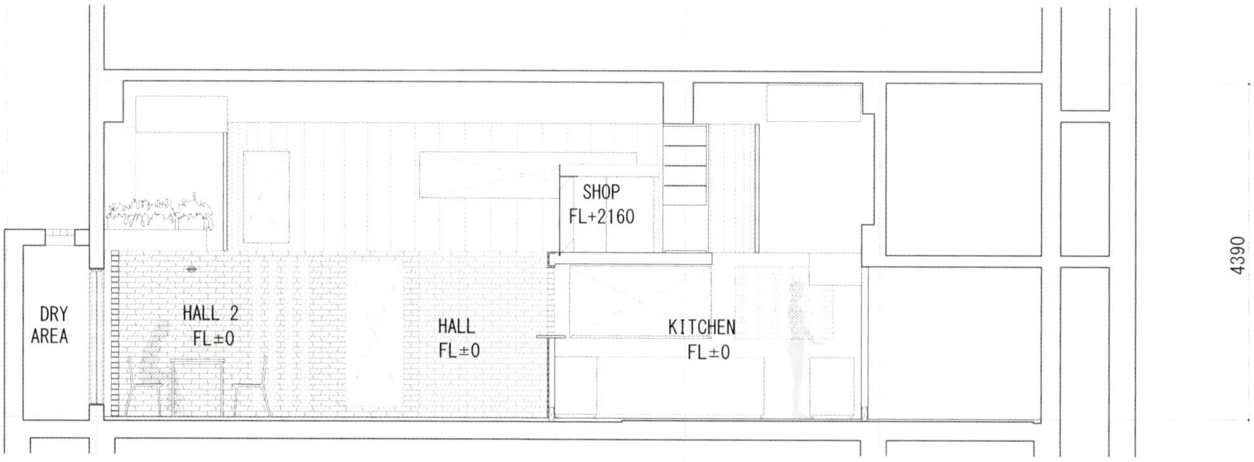

In upper space, white wall hides original wall and equipment like air conditioning system or piping. And the wall can serve as frames which show some graphic, plants and scenery of the restaurant. In addition, it surrounds stairs space and make big void. In lower space, you can slightly feel the border (plan) of white wall above your head and it's quite different from the plan of the restaurant floor partitioned by brick wall. The brick wall turns in right angle many times and makes hall space on the inside. And it also makes private room, kitchen and storeroom on the outside. The hall space is divided into three places.

上层空间的白色墙面遮挡了原建筑的墙面以及空调和管道等设备器材。这里的墙面也能作为展示墙，摆放一些画、植物和餐厅布景。另外，环绕楼梯间的空间更加宽敞。在楼下，你会感受到顶上的白墙和餐厅楼面砖墙的区别。砖墙在大厅内侧构成具有多个直角的空间，并在外侧构成包间、厨房和储藏室。大厅空间被分为三个区域。

Salon des Saluts 酒馆

Project > 项目名称 > Salon des Saluts

Architect > 设计公司 > Sinato 设计公司 (Chikara Ohno、Masaki Ito)

Location > 项目地点 > 日本东京

Built area > 项目面积 > 53.51 m²

Photographer > 摄影师 > Toshiyuki Yano

项目介绍 PROJECT DESCRIPTION

Designers designed a small Wine Bistro fronted on Sei-jou-ki street, Nishiazabu. Our client requested two things, one is concept keywords 'Nature' or 'Earth' and another, 'Openness', so anyone can look at interior of the shop from outside. Designers were afraid that the flat glass skin along the building outface makes the contrary stronger on a boundary between the inside and outside.

这家小酒馆位临西麻布的一条繁华街道。客户有两条要求：一是设计的关键词为"自然"或"土地"；二是"开放"，让所有人都能从外面看到店内的情况。以前设计师担心在建筑的外表面用平板玻璃会使内外对比过于强烈，现在看来效果还不错。

For the connection between the outside street and inside space, considering more worthy of ambiguous and deep transparency, as a result they decided to adopt a way crowding an opening of the building with small glass boxes running across over the outface.

在室外街道和内部空间的连接处，设计要具有适当的模糊性与深度通透感，因此在建筑物门口的外立面堆砌很多小玻璃箱。

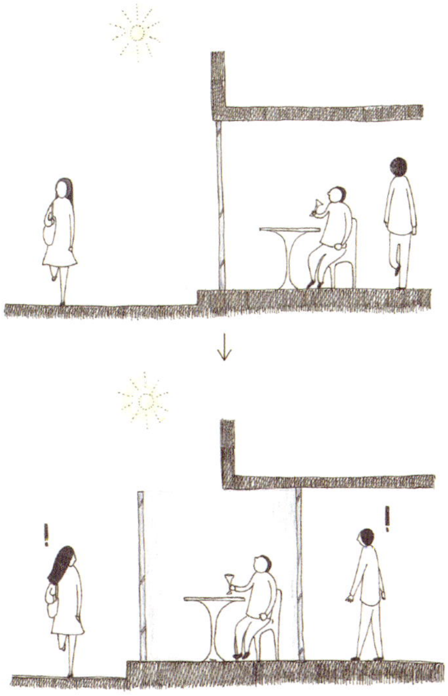

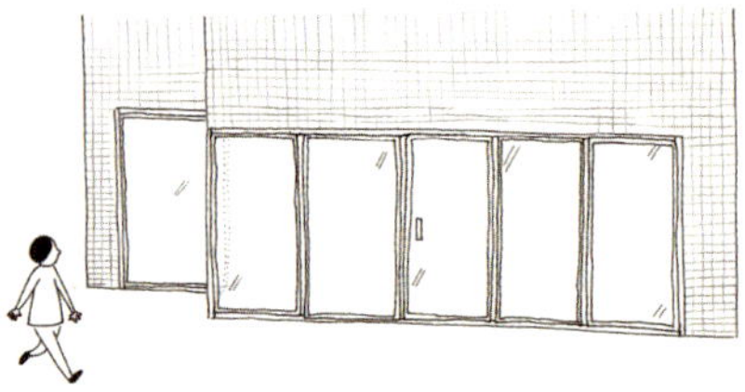

The flat glass skin along by the building outface

↓

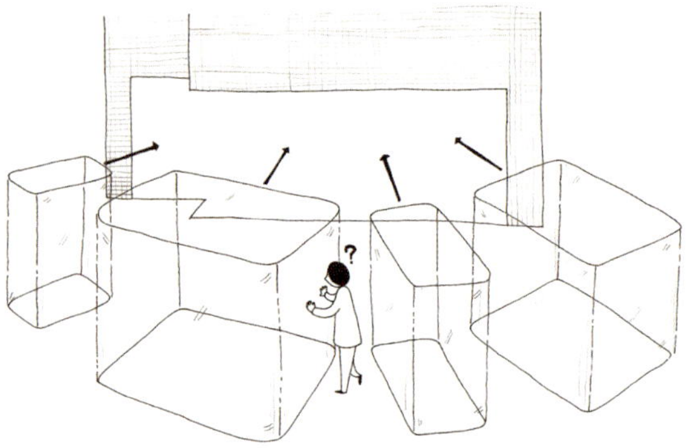

An each corner of box processed a curved surface.

↓

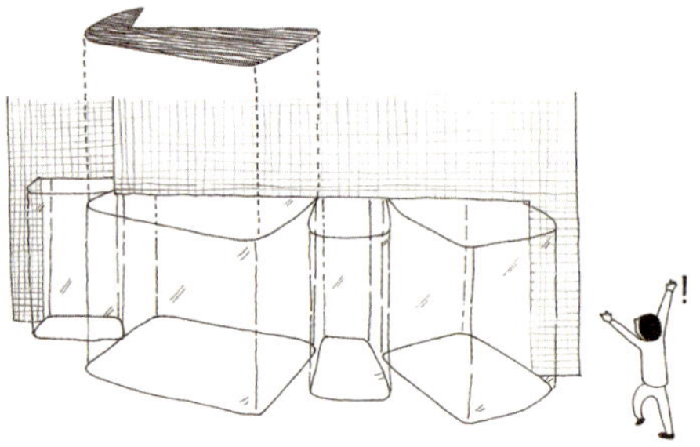

Crowding an opening of the building with small glass boxes running across over the outface and inside space put a roof.

↓

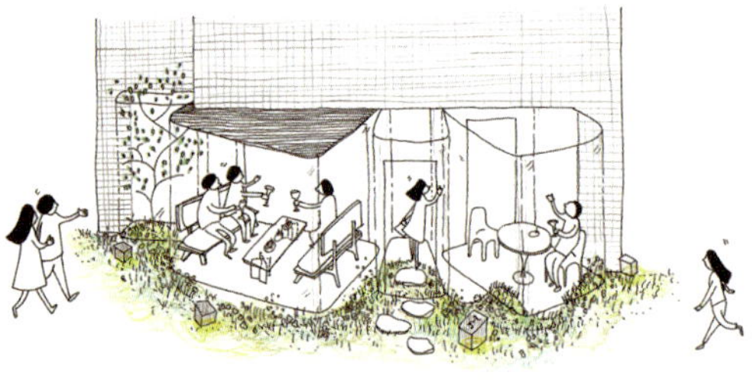

Plants run free curve with floor tile and crosses over the glass wall swung as like shore and wave.

Except one box served for private room in inside, others are served for a court with single olive tree with approach and terrace are outside, so it's difficult to know which is a real boundary between in and out because this line runs very complicatedly. Besides, an each corner of boxes processed a curved surface brings the visually distortion and blurred transparency to the view from the outside.

Another important component is that the grand plants filled with asparagus, rosemary, ivy and etc. running free curve around floor tile and crossing over the glass wall, swung as like shore and wave. It emphasizes the ambiguousness of the boundary.

除一个玻璃箱作为内部的客房使用外，其他玻璃箱里都栽种了橄榄油树，通道和露台都在玻璃箱外面，这样就很难找到内外空间的真正界限了，因为行径路线十分曲折。此外，箱体的每个角均被弯曲成曲面，使人从外面看进来会产生视觉变形和模糊感。

另一个重要设计是遍布店内的芦笋、迷迭香、常春藤等大型观赏植物。它们在地板砖上盘旋，像海浪一样跨过玻璃墙，更隐晦了室内外边界。

Designers hope what they designed as shop front becomes to be a kind of public amenity over own character and makes a good experience for not only customer, but for people walking along the street and the city environment itself.

餐厅的门面成为一道独具匠心的公共设施，不仅为顾客，也为经过该街道的人们和城市本身制造一种愉悦的环境。

Nevy 餐厅

Project > 项目名称 > Nevy 餐厅

Architect > 设计公司 > Concrete Architectural Associates 设计公司

Designer > 设计师 > Rob Wagemans, Janpaul scholtmeijer, Jari van lieshout, Erik van Dillen

Built area > 项目面积 > 220 m^2

Photographer > 摄影师 > Ewout Huibers

concrete

项目介绍 PROJECT DESCRIPTION

concrete was asked to redesign the formerly known restaurant Onassis. The restaurant is a new member in the succession of two existing restaurants; delicacy restaurant "envy" and wine bar "vyne". The facility for this location is a fish restaurant called "nevy". To create a visual separation between the two parts black and white are chosen. In the white part, the ceiling and floor are black. The opposite appears in the black area. There you see a white floor and ceiling.

Concrete 公司负责重新设计以前广为人知的 Onassis 餐厅。这家新的餐厅是由 envy 餐厅和 vyne 酒吧合并而成，新餐厅是这家叫"Nevy"的海鲜餐厅。为了使两部分在视觉上有所区别，设计师将两部分分别设计为以黑色和白色为主色调的空间。白色部分的天花板和地板是黑色，黑色部分则相反，地板和天花板均为白色。

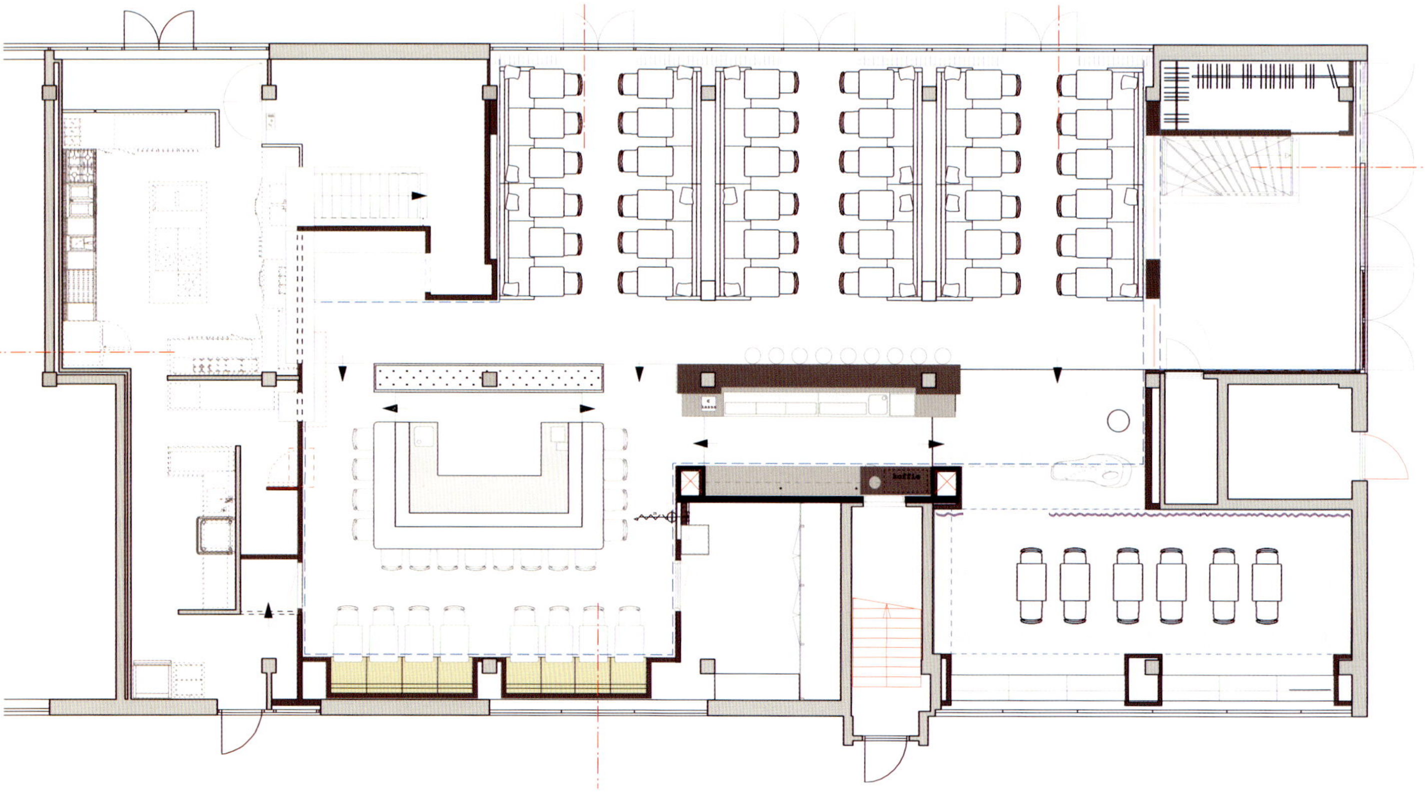

In this part of the restaurant you can find the best view of the "Ij" and Amsterdam north. Concrete chose to place six couches, white marble tables and white high gloss revolt chairs cross-grained the façade. In this way guests can optimally enjoy the view. The couches have a "deeper" seating so that the customers can relax longer after finishing dinner.

在白色区域可从最好的角度欣赏到阿姆斯特丹的美景。Concrete 设计公司在这里布置了六张长沙发、汉白玉大理石材质的桌子和表面带有斜纹的白色旋转椅。在这里客人能在舒适的环境中欣赏美景。沙发非常柔软，可供客人在晚餐后躺在这里小憩。

In the back of the restaurant you can find the raw bar. In this area the chef is the main focus point. The raw bar experience is different compared to the white restaurant; the walls are made of black marble, the floor of white marble and the ceiling is painted white. Besides the raw bar this area also functions as the distribution point of the kitchen and can be seen from the entrance of the restaurant.

黑色区域放有生鲜类食品自助柜台，厨师是这个区域的主角。自助柜台与白色区域截然不同。这个区域的墙面是黑色大理石，地板是白色大理石，天花板同样也被粉刷成白色。自助柜台旁边的区域是厨房的另外一个分配点，在餐厅的入口处就可看到。

The entrance contains a large visual of an X-ray photographed fish. The visual refers back to the visual identity in which the fish can be seen in various variations. The wardrobe and the stairs to the cellar where the toilets are situated are located at the entrance. Direct at the entrance you step into the white restaurant. On your left is a doorway that leads to the wine cabinet. This space also provides room for private dining for about 20 people.The wine cabinet area can be separated by a velvet blue "yves klein" curtain. This colour is part of the visual identity of nevy.

入口处摆放了一幅鱼的 X 光相片，这幅鱼的非常态下的视觉效果图成为了餐厅的标志。衣柜和通往地下室的楼梯也位于入口处，卫生间位于地下室。餐厅的白色区域直接对着入口，左边的通道可直达酒柜。这个区域也有一个可供 20 个人就餐的包间。酒柜区域用一个天鹅绒蓝的布帘隔开，布帘的色彩也是 Nevy 餐厅的标志。

“珍珠和鱼子酱”餐厅

Project > 项目名称 > “珍珠和鱼子酱”餐厅

Architect > 设计公司 > Concrete Architectural Associates 设计公司

Location > 项目地点 > 阿联酋

Built area > 项目面积 > 1430 m^2

Photographer > 摄影师 > Richard Thorn

项目介绍 PROJECT DESCRIPTION

"Pearls & Caviar" is a restaurant on the site of the Shangri-La hotel in Abu Dhabi. It is placed in a solitaire building on the waterside of the river. The client asked for a contemporary restaurant, which nevertheless represents the origins of the very region. "Pearls & Caviar" represents the new Arabian lifestyle, a luxury fusion of east and west, black and white, occident and orient, light and shadow, the extrovert and the introvert, the intimacy and the view.

"珍珠和鱼子酱"餐厅位于阿布扎比香格里拉大酒店旁边一座临河的独立建筑中。客户要求现代的设计风格，并且要带有当地的地域特色。"珍珠和鱼子酱"餐厅引领着阿拉伯新的生活方式，是东西方、黑白色调、欧美与东方文化、光影效果、外部与内部以及私密空间与风景的广泛融合。

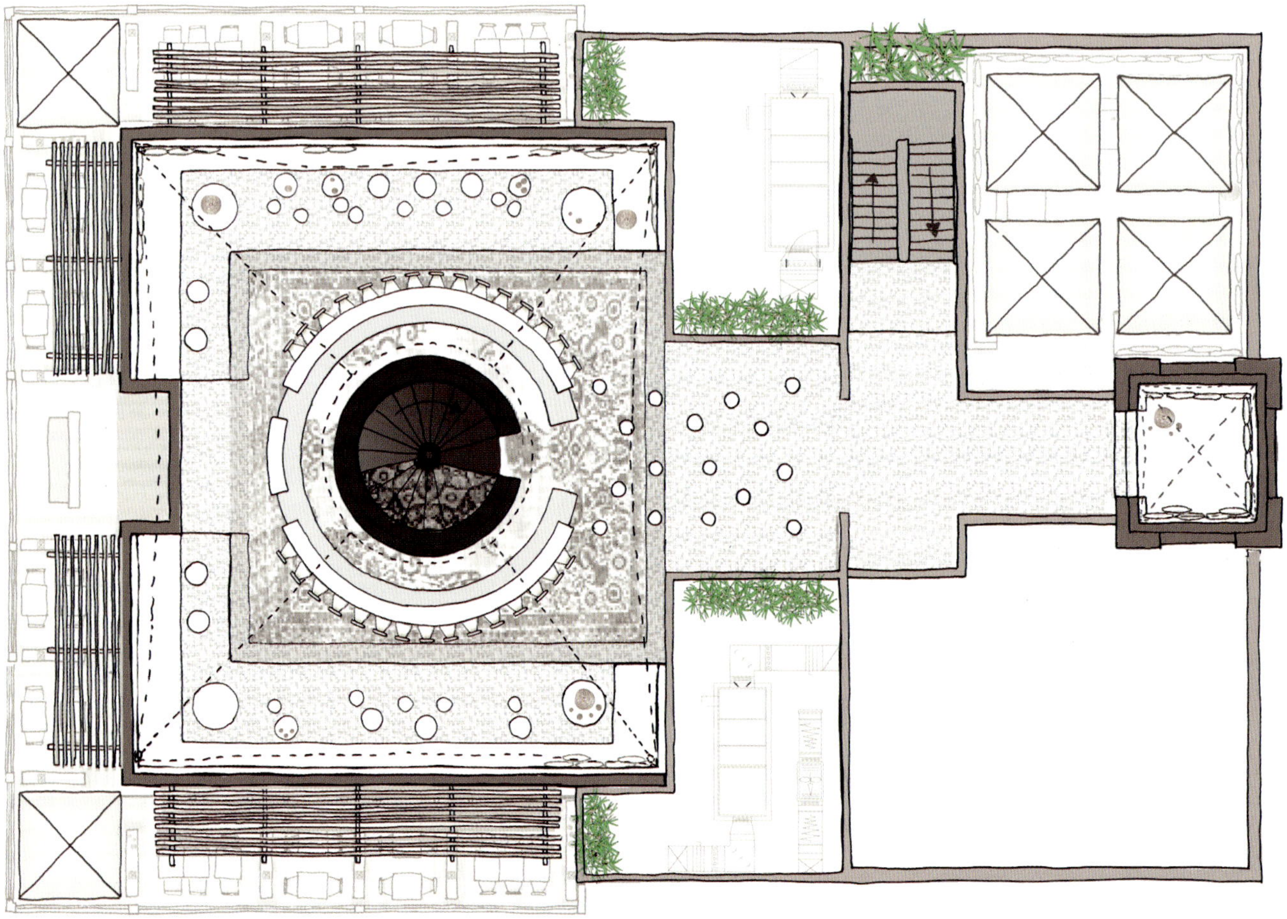

pearls & caviar

The basic idea of the design of the restaurant has been to create an abstraction of the commonly used oriental forms and materials without loosing their richness. To achieve this amount and the richness of the oriental patterns and forms which are found in traditional oriental spaces are kept but without color. All colors are replaced by either shades of black or shades of white — both in combination with silver. Therefore the restaurant is divided in two parts, which also gives its name: "Pearls" and "Caviar".

设计的基本思路是使用一般的东方元素和材料，同时又不失华美。为达到这种效果，传统的空间里布置了大量丰富的具有东方文化特色的模型。这些模型不是彩色的，主色为黑色、白色和银色。餐厅被分为两部分，也因此被命名为"珍珠和鱼子酱"。

The "pearl" part is represented in the terrace on level 1. The terrace is divided in 3 subspaces: the main terrace and two private terraces, which can be used separately. The pearl areas are extrovert spaces composed completely in shades of white and silver: The flooring is a big oriental carpet made of glass mosaic tiles in the white-silver color range. A circular bar made of polished stainless steel with a bar top of white marble is placed around the tower of the main terrace. White upholstered, deep lounge benches are placed along all borders of the single terraces and white sails cover each of them as sun protection and light diffuser.

命名为"珍珠"的部分位于一层的露台上。露台又分为三个子空间：主区和两个私人空间。每个空间都能单独使用。这个区域是对外开放的空间，区域的主色调为白色和银色。地板上铺着具有东方特色的大地毯，玻璃马赛克闪着银白色的光。圆形的吧台用抛光不锈钢做成，表面是白色大理石。每个单独露台的角上都摆放着放了白色软垫的长椅，上面覆盖着帆布，保护其不受阳光直射。

The "Caviar restaurant" is situated on level 0. It is an introvert and intimate space in the inside of the building composed completely in shades of black and silver: The flooring is a big oriental carpet made of glass mosaic tiles in the black-silver color range. The restaurant is divided in little intimate "chambers" on the rim of the carpet separated by ball chain curtains and side tables made of wengé. On top of them Persian accessories are placed which function as light diffusers. Around the central tower a circular bench is placed upholstered in black leather. The back parts of the benches stretch over the full height of the wall of the tower. And a straight bar in stainless steel with black marble bar top.

命名为"鱼子酱"的部分在底层。这里是大楼的内部隐蔽空间，主色调为黑色和银色。地板上铺着具有东方特色的大地毯，玻璃马赛克的颜色是黑银色。这里沿着地毯的边缘分为一个个小的"内庭"，用珠帘隔开。靠墙的桌子用黑铁木做成，上面摆放着波斯风格的装饰品，用来反射光线。中心塔楼的周围摆放着黑色皮革软装长凳，它的颜色映射到塔边的整个墙面上。直线型的吧台用不锈钢做成，表面覆以黑色大理石。

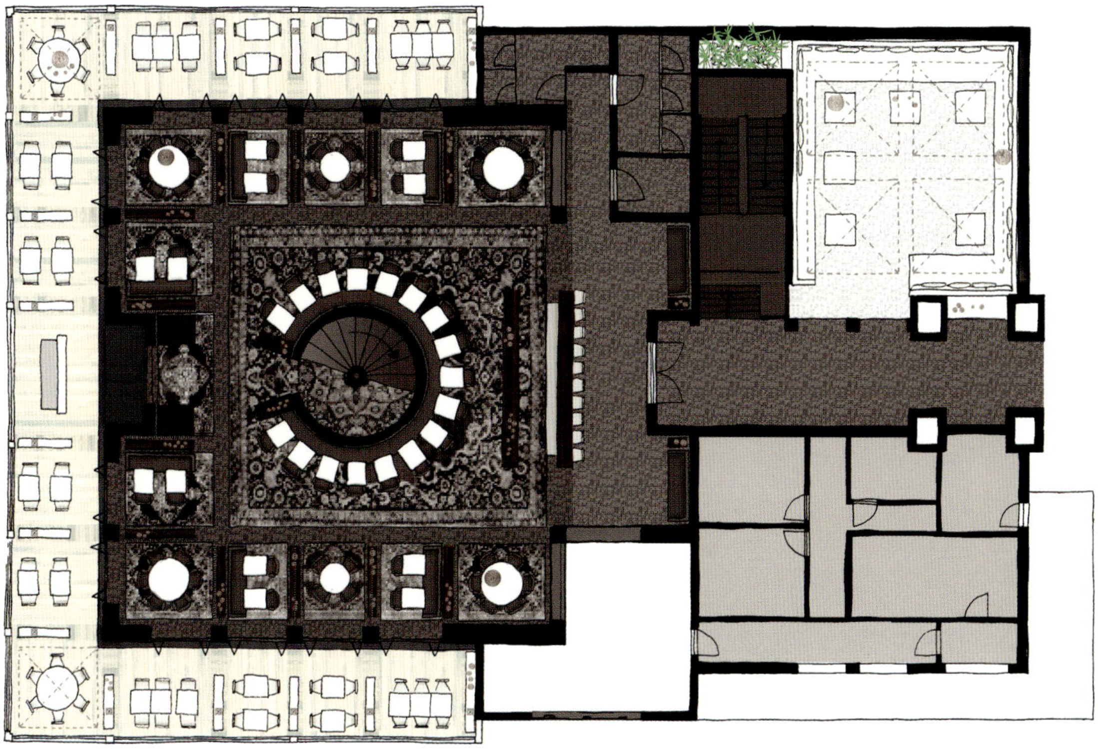

国民餐厅

Project > 项目名称 > 国民餐厅

Architect > 设计公司 > Rockwell Group

Location > 项目地点 > 美国纽约

Built area > 项目面积 > 623.1 m^2

Photographer > 摄影师 > Frank Oudeman

项目介绍 PROJECT DESCRIPTION

Looking to elegant bistros and cafes in Paris, London and New York for inspiration, Rockwell Group designed three distinct areas on the main floor of the restaurant, creating the appearance of an established restaurant built over time. This refined sensibility will meld with the grandeur of the restored Emery Roth building that houses Benjamin Hotel.

参考了巴黎、伦敦、纽约具有优雅风范的小餐馆和咖啡厅后，罗克威尔设计公司构思了设计思路。他们将餐厅的主要楼层分为三个区域，餐厅的外观是做旧的。精细的设计与同在本杰明酒店的经修复的 Emery Roth 大楼壮丽的外观相融合。

While the bar lounge, main dining room, and open kitchen all have distinct personalities, a signature dark blue/teal-colored leather was created specifically for this restaurant to draw all the spaces together. This leather used throughout the restaurant on banquettes and chairs introduces a vibrant and welcoming flourish to the lush, sophisticated environment.

吧台休息区、主要就餐区和开放式厨房都具有与众不同的设计特点，富有特色的深蓝色和青色皮革是餐厅的主要材料，使整个空间有机统一。它主要用在人行走道和椅子上，为豪华、精细的环境增添了活泼气氛，也彰显了迎客之道。

An oiled bronze curved staircase leads guests from the restaurant to the second floor, where the intimate library, private dining rooms and pre-function area are housed, as well as the residence lounge specifically designed to serve hotel guests. The library has a bespoke hutch that opens for bar service. The seating in this area can transform from elegant lounging to private dining for up to 28. Its corner location provides guests with a view uptown on the bustling Lexington Avenue.

一架被刷成青铜色的弯曲楼梯通向二层，客人可从这里到达私人图书馆、私人晚宴室和其他一些功能区。休息厅是特意为酒店的客人设计的。图书馆里有为酒吧服务定做的储藏柜。从休息厅到私人用餐区的座位有 28 个，由于位于拐角处，客人坐在这里可以欣赏到繁华的莱克星顿大街景色。

To add a contemporary and casual spin to a traditional design element, the wood paneling wrapping around the restaurant is washed with a cream colored lacquer. This color also appears in the tiles for the restaurant and the private restrooms.

为使传统的设计元素充满现代感和休闲气息，餐厅周围装上了刷上奶油色漆的木制护墙板。餐厅的瓷砖和私人休息室也采用了这种颜色。

In the bar lounge area are cerused oak floors, classic Thonet bar stools, and a custom light fixture made of faceted antique mirror and filament light bulbs that defines the ceiling over the bar. The main dining room and kitchen café both have glazed cement tile floors, each with a distinct classic pattern. The open kitchen is framed by bronze and seeded glass screens.

吧台休息区是白橡木地板，摆放了古典奈特高脚凳，定做的灯具上嵌有许多古董镜子，一个个灯泡装饰了吧台上面的天花板。主要就餐区、厨房和咖啡厅使用光滑的水泥花砖铺装，每块上面都有古典花纹。开放式厨房用古铜色的玻璃屏风分隔。

纽约休闲餐厅

Project > 项目名称 > 纽约休闲餐厅

Architect > 设计公司 > Nemaworkshop 设计公司

Location > 项目地点 > 美国纽约

项目介绍 PROJECT DESCRIPTION

For the design of Delicatessen, nemaworkshop explored the concept of urban identity, namely the vibrant SoHo neighborhood and more specifically the NY newsstand. Typically overlooked because of its ubiquity, the newsstand serves as an atypical yet appropriate model.

纽约休闲餐厅是对都市风格的探索，是喧闹的 SoHo 社区的缩影，具有浓厚的美国味，它很普通，甚至容易被人忽视，却是一个十分符合当地风格的非典型案例。

DELICATESSEN

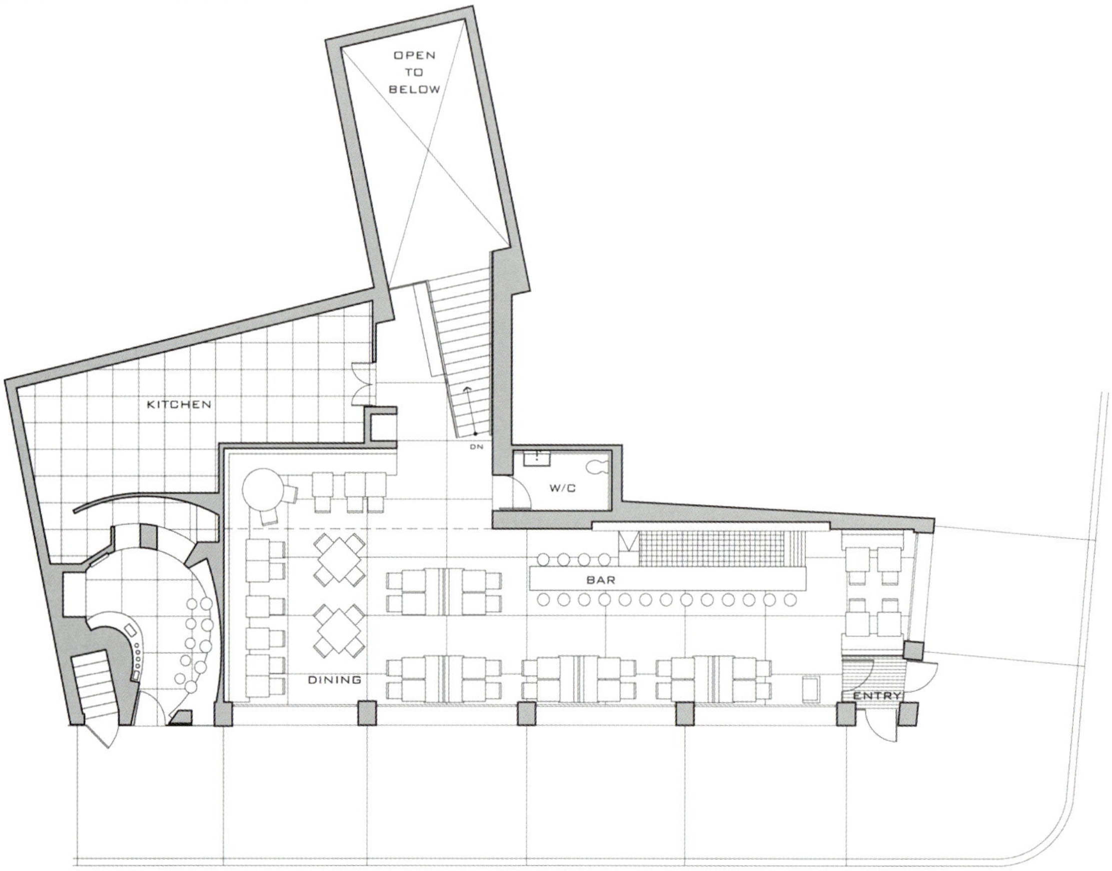

PLAN: STREET LEVEL

MASTA KILLA

DELICATESSEN

The concept of informal urbanity develops complexity as one travels downstairs into the more intimate spaces below street level. The existing boiler room has been transformed into a glass-roofed lounge lined with an oversized mural by local artist Juan Jose Heredia.

纽约休闲餐厅用下沉式的入口来诠释洒脱的都市风格，远离尘嚣，使人倍感亲切。以前的锅炉房被改装成一间玻璃屋顶的休闲厅，墙上装点着纽约艺人的壁画的放大版。

They are open, in-and-out thoroughfares which foster a unique breed of social energy. In order to capture this energy, the main dining space opens up to the bustling street corner with large retractable steel-and-glass garage doors. Traditional boundaries between street and restaurant evaporate, allowing the main dining space to be loosely defined and astonishingly inviting. Underneath the bar stands a collection of clear soda pop bottles while behind the bar the wall is clad in leather subway tiles.

良好的开放和通透感为社交活动提供了很好的条件，为了获得这种便利，餐厅正面对着繁闹的街市敞开，只有一扇可伸缩的铁制与玻璃的车库门相隔。这样，街道与餐厅的传统边界消失了，餐厅的布置显得松散而具有吸引力，吧台下面排列着很多苏打瓶收藏品，吧台后面的墙面由皮革地铁贴砖装饰。

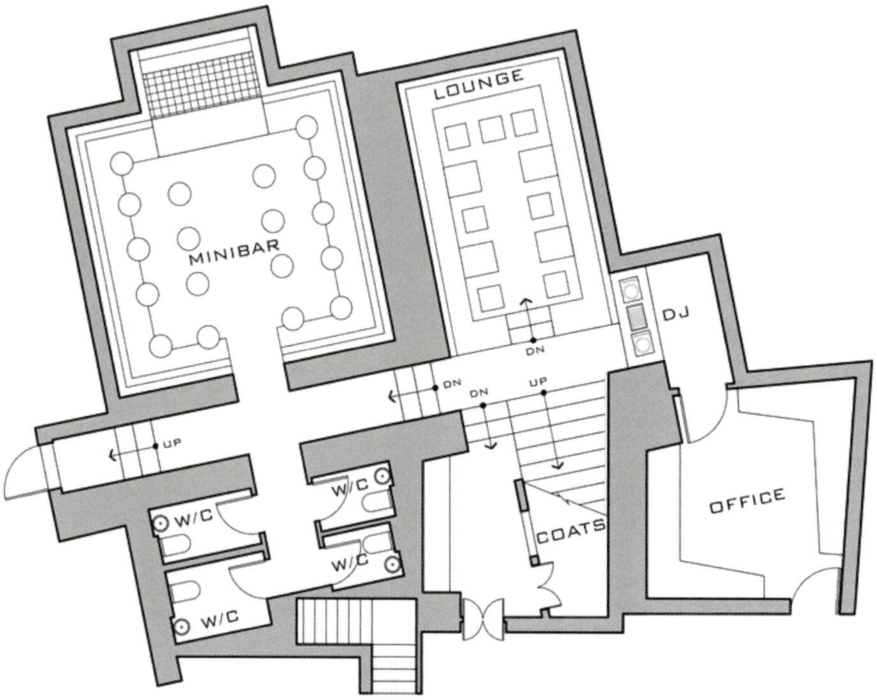

PLAN: BASEMENT LEVEL

Just a few steps down from the lounge / courtyard is a minibar. As the name implies, it's a cozy underground bar where the walls are lined with vintage mini-bottles. Ultimately, the design of Delicatessen from the multi-leveled spatial organization to the comfortable yet sophisticated materiality is rooted in the concept that the space should be both inviting and accessible.

穿过休闲厅就来到了一间迷你吧，这个舒适的地下酒吧的四壁摆放着复古式样的迷你酒瓶。该案用多层次空间的设计、繁复而舒适的材料创造出了一个引人入胜而亲切可人的空间。

Cho Gao 概念餐厅

Project > 项目名称 > Cho Gao 概念餐厅

Architect > 设计公司 > Creneau 国际

Location > 项目地点 > 澳大利亚墨尔本

Photographer > 摄影师 > Philippe Van Gelooven

Creneau International
ideas of the anti-ordinary

项目介绍 PROJECT DESCRIPTION

Creneau International set off for Melbourne (Australia) to open the first restaurant in its new Cho Gao franchise – a concept that puts genuine Asian cuisine firmly on the menu. Forget your run-of-the-mill, westernized eateries, like takeaway Chinese. Instead, get ready to enjoy original Oriental dishes that will get your taste buds tingling, served with charm in an authentic décor.

Creneau 国际在澳大利亚墨尔本开了他们第一家 Cho Gao 餐厅，这是一家主营亚洲美食的餐厅。与一般的外卖不同，在这里，客人可以品尝到最地道的东方美食，感受美食的魅力。这间式样新颖、现代的餐厅是东方艺术在餐饮上的体现。

This timeless and stylish new restaurant concept is just the answer to kitschy Oriental eating-places. "The latest trends all seem to crop up first in the Far East" says Werner Vanherle, CEO of Creneau International, about the inspiration behind Cho Gao. "Asia is ultra-trendy. At the same time, you simply cannot ignore the sheer wealth of age-old traditions that Asia has to offer. It's a wonderful melting pot of authenticity, energy and creativity and an enormous source of inspiration for us."

永恒与时尚新概念是东方饮食场所的最佳诠释。Creneau 国际的 CEO Werner 表示，开办 Cho Gao 餐厅的原因在于最近的流行趋势都首先出现在东方，亚洲是潮流的引领者。同时设计师也不能忽视亚洲悠久的传统文化，这是一笔巨大的财富。它能够激发设计灵感，提升创造力，是精神鼓舞的源泉。

WI
2000

Anne-Mie Stevens, interior designer at Creneau International, is the creative brain behind Cho Gao. For her, it all began with two Oriental items she snapped up in an antiques shop. An old wooden panel and a little bench formed the basis for all the other furniture.

Anne-Mie Stevens 是 Creneau 国际的设计师，是 Cho Gao 餐厅的主要设计者，她对于东方设计元素的构思源于在古董店买的古董。一块古老的木板和一些长凳是餐厅家具的核心。

The concept for the interior is a successful balance between traditional soberness and trendy design. The restaurant's soul consists of a mixture of unpretentious fixtures and fittings whose very simplicity effortlessly succeeds in catching the eye and drawing attention. And consciously using recycled teak for features such as the room dividers and benches gives the whole place a lived-in look that comes across as totally authentic.

内部设计的构思是传统设计与潮流设计的完美结合。餐厅的核心在于内部装修和家具陈设，虽然朴素平实，但却轻易成功地吸引了视线，引起了客人的注意。使用可回收的柚木是设计的一大特色，屏风和长凳也使用这一材料，给人一种这里长期有人居住的感觉，更为亲切和真实。

The people who come to Cho Gao not only get Asia on their plate, but are also immersed in the typical atmosphere of the Far East," says the designer. "We wanted to create something understated and authentic, with a mixture of sturdy pieces and refined accessories.

到Cho Gao餐厅就餐的客人不仅能品尝到东方美食，同时也融入到一个具有典型东方特色的环境中。设计师想打造一个朴素平实而真实亲切的环境，同时家具要牢固耐用且不失文雅。

中心酒廊寿司吧

Project > 项目名称 > 中心酒廊寿司吧
Architect > 设计公司 > Bluarch 设计公司
Location > 项目地点 > 美国

项目介绍 PROJECT DESCRIPTION

Central Lounge and Sushi Bar is a supper club, a hospitality venue offering both food and music. The design approach was to offer the indulgent environment of a lounge and the sensual experience of the finest sushi in a rich and eclectic space.

中心酒廊寿司吧是一家美食俱乐部，美食和音乐共享其间。设计师为休闲娱乐与上乘的寿司美食提供了一个宽松的环境，是一个具有丰富体验和折衷主义韵味的空间。

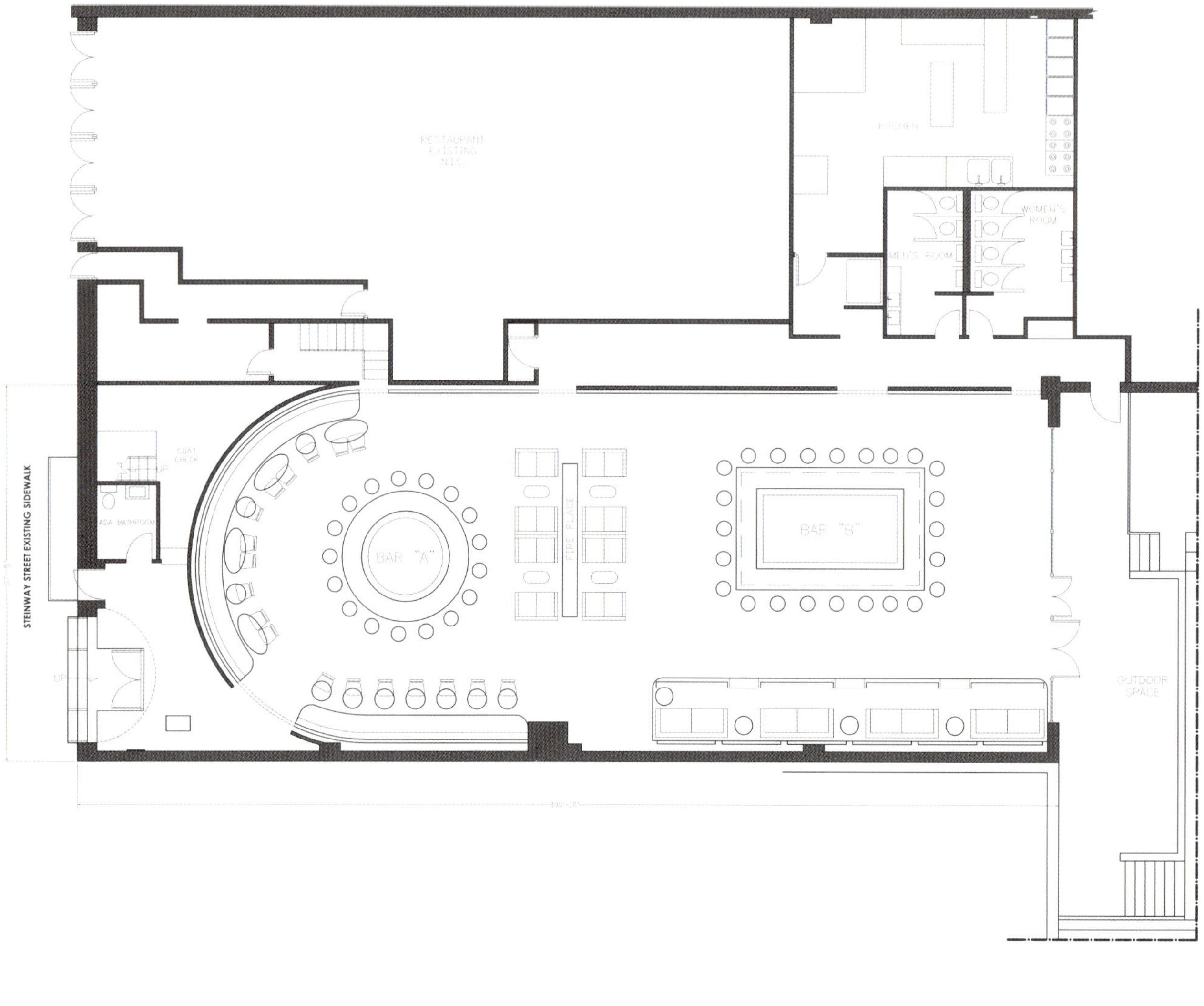

BACARDI

The venue has two bars, one rectangular and one circular. Both are clad in tufted white upholstery and are topped with slabs of back-lit onyx. Above the rectangular bar, five walnut shelters curve at different heights to create a refined sense of intimacy.

该场地有两个不同形状的酒吧，一个是矩形的、一个是圆形的。两者都是由白色簇绒织物和背投照明的玛瑙墙装饰而成。在矩形吧台之上，高低不同的五个胡桃木柜子，如波浪般起伏，巧妙地形成了一种私密感。

Each shelter has a cut-out area showing a black mirror and a Swarovski crystal sconce.

每个柜子都有一处凹进的面积，上面安装了黑色镜面和施华洛世奇水晶壁灯架。

The circular section of the venue is surmounted by a mezzanine which opens into a double height topped with ivy and LED lights. The perimeter of the round area is delineated by a round banquette above which round upholstered panels in different sizes line the walls. The round mezzanine above visually connects the state-of-the-art DJ booth with the patrons below and ensures an always vibrant atmosphere.

该餐厅的圆形区域被夹层覆盖，两层楼高的天花板上装饰着常春藤和 LED 灯。圆形区域的周围是环形走廊，许多圆形装饰板装点于侧壁。圆形夹层从视觉上将最先进的 DJ 台同楼下的泡吧客人联系起来，让整个空间充满活力。

Angelini 意大利酒吧餐厅

Project > 项目名称 > Angelini 意大利酒吧餐厅

Architect > 设计公司 > BCV 建筑设计公司

Location > 项目地点 > 泰国曼谷

Built area > 项目面积 > 845 m^2

项目介绍 PROJECT DESCRIPTION

Angelini is a reinvention of an existing fine-dining Italian restaurant and bar within a five-star resort hotel on the banks of Bangkok's Chao Phraya River. Central to the new design is the relocation of its main entry to the hotel's lower level to create connection to other venues at the riverside level, and to distinguish Angelini as its own destination.

Angelini 意大利酒吧餐厅位于泰国湄南河边的香格里拉五星级酒店，它建造的重点是让酒店底层的大厅与湄南河边的休闲区自然过渡，让餐厅引人入胜。

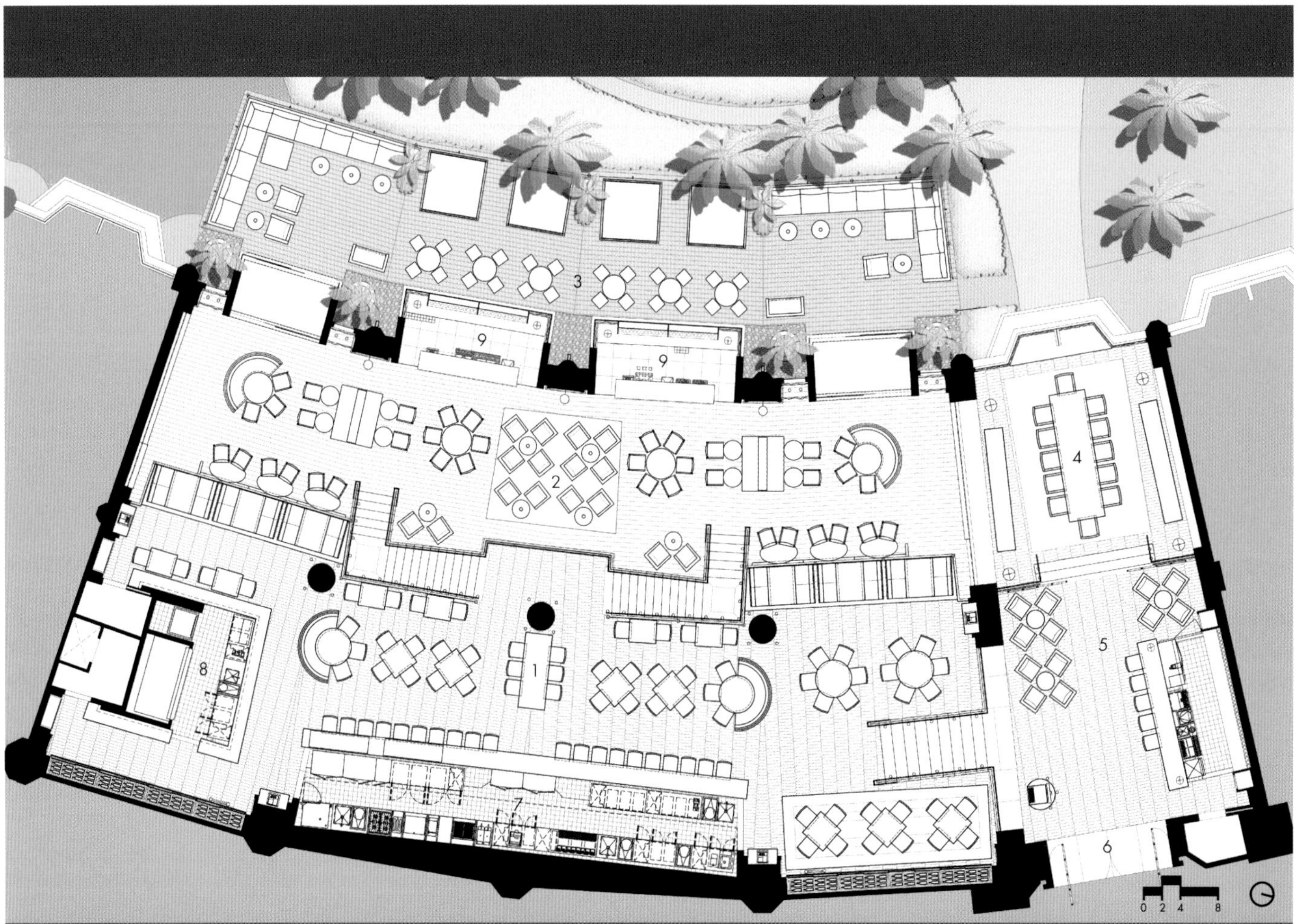

1 Cafe Dining **2** Lower Level Lounge **3** Lounge Terrace **4** Private Dining **5** Bar/Lounge **6** Lobby Level Entry **7** Exhibition Kitchen **8** Pizza Kitchen **9** Bar

The lower level is transformed into a new bar and lounge space, with curved seating arrangements allowing views to a live performance stage beneath the upper balcony. A series of dramatic new window bays are inserted between existing columns to define the newly important riverside façade of the hotel. The two most central of these window bays house striking twin bars, with soaring decorative bottle displays. The outermost window boxes serve as vestibules to a sweeping exterior wood deck terrace. Set beneath a glass and steel canopy, this exterior lounge is a lush outdoor room connecting Angelini to poolside gardens.

The existing architecture has been reconceived in a bolder, modern expression that plays upon the contrast of light material against dark, translucent against opaque, polished against honed, slick against organic, with cool highlights set against the deep warm tones of the project. Marble floors at the upper level contrast with the deep hues of the lower lounge wood floors and walls, the two levels blended warmly with veneer wood paneling between historic ceiling beams. This contrast of dark and light has its best expression in glass tiled column wraps at the central columns. Illuminated decorative panels and light fixtures create bold strokes of amber light that bathe the restaurant in a golden hue for a warm atmospheric effect.

酒店的底层被改造成一个全新的酒吧和休息室，曲线的座椅排列让楼下的客人可眺望到楼上的包厢。大厅柱子间的一扇扇大飘窗尽收酒店外的河岸美景，同时也形成了美丽的酒店外立面。最中间的两个飘窗由许多瓶子装饰出两个相同的吧台。最外面的窗栏结合宽阔的木质露台形成门廊。钢架和玻璃搭成了室外休闲厅的天棚，这里成为连接 Angelini 餐厅和泳池花园的地方。

该空间突出明与暗、虚与实、高光与亚光、平滑与层次的对比，用冷调的浅色光投射暖调的深色建筑，二楼的大理石地板与一楼的深色木地板和墙面形成对比，而顶上的木质胶合板将两层楼的基调很好地糅合在一起，大厅中间柱子上的玻璃砖也有明暗的对比。照明嵌板和灯具发出琥珀色的灯光，将大厅染成温暖的金色，尊贵与和谐不言而喻。

La Espiga 餐吧

Project > 项目名称 > La Espiga 餐吧

Architect > 设计公司 > CT Arquitecto Asesor 设计公司 (Carlos Tejada.)

Location > 项目地点 > 西班牙

Built area > 项目面积 > 387 m²

Photographer > 摄影师 > Enrique Satrústegui、Carlos Tejada.

ARQUITECTO ASESOR

项目介绍 PROJECT DESCRIPTION

Since 1929, "La Espiga" is a gastronomic meeting space of reference in San Sebastián-Donostia. The building where it is located is in a reconstruction phase. Its ground and underground floors will house the new premises that are given a renewed contemporaneous image, projecting the future ahead without loosing contact with tradition of the past. A place dedicated to the usual customers and to the city visitors, where they would feel identified and enjoy agreeable moments. A warm and timeless place inspired by the landscape, culture, light and materials that characterizing Guipúzcoa, Donosita. It escapes a fashionable design. The ground floor is a tribute to the province, the underground floor to the city and loyal customers.

自 1929 年建成以来，La Espiga 餐吧一直是西班牙圣塞巴斯蒂安城的一家集美食与社交为一体的餐厅。该餐厅正在进行重建，改建后，其底层和地下一层将成为此餐厅新的经营场所。设计在保持传统特性的基础上兼具现代感。设计师希望来到此地的游客能度过惬意的时光。因其具有地方特色的风景和文化，以及独特的灯光和材料搭配，使其成为一处能感受到温暖和永恒的绝佳去处，绝非单纯的跟风。该建筑的底层供省市要员使用，地下一层供游客和尊贵的食客使用。

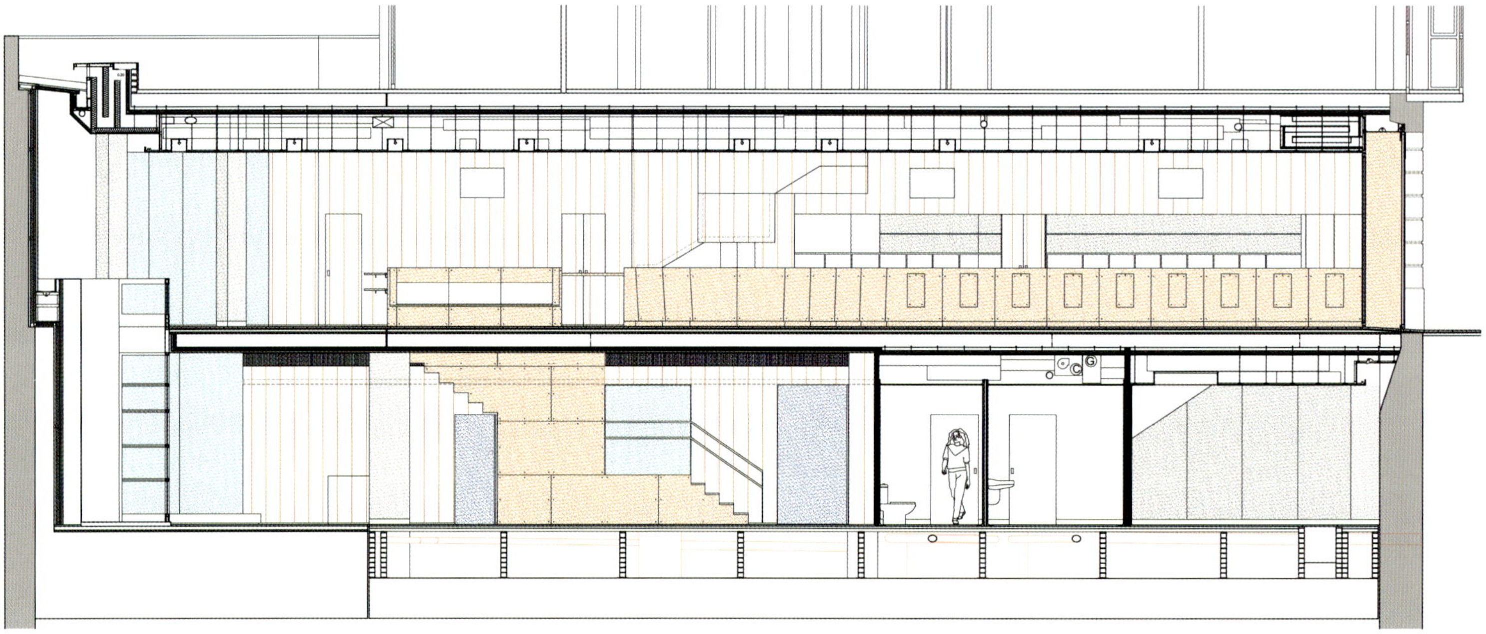

LONGITUDINAL SECTION

LA ESPIGA
GRAN BAR
48

Six mural reproductions of mural painter José María Sert, in San Telmo: wise land, owners land , sailing land, the altar of race, holy land ant forgers land, symbolizing the six rivers that compose the Guipúzcoa territory: Bisasoa, Oyarzun, Urumea, Oria, Urolay, Deva rivers.

墙上挂着六幅由圣太摩壁画大师制成的 José María Sert 的壁画复制品：《智慧之地》、《业主之地》、《航海之地》、《民族祭坛》、《圣地》、和《铁匠之地》，象征组成 Guipúzcoa 区域的六条河流：Bisasoa、Oyarzun、Urumea、Oria、Urolay、Deva 河。

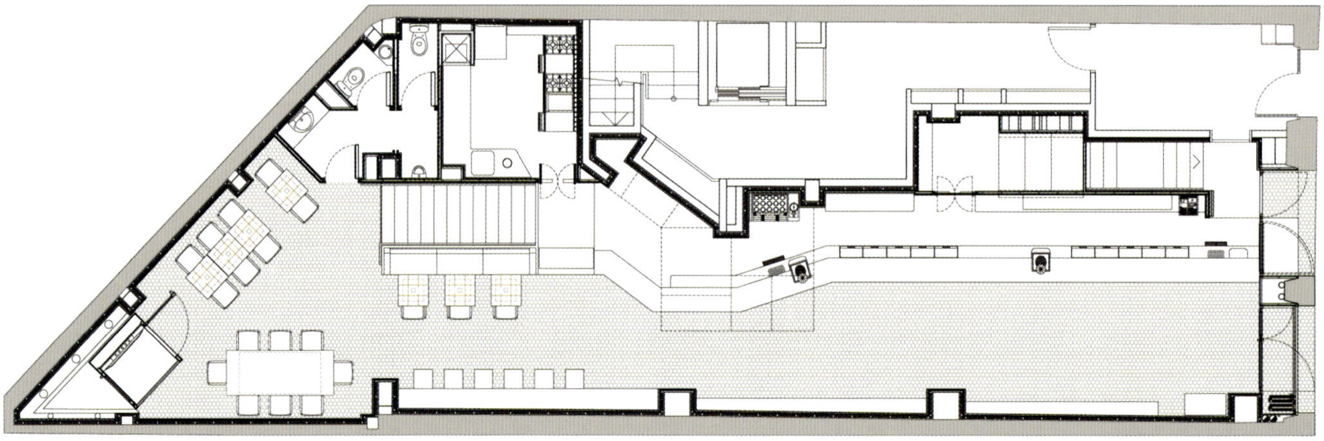

GROUND FLOOR

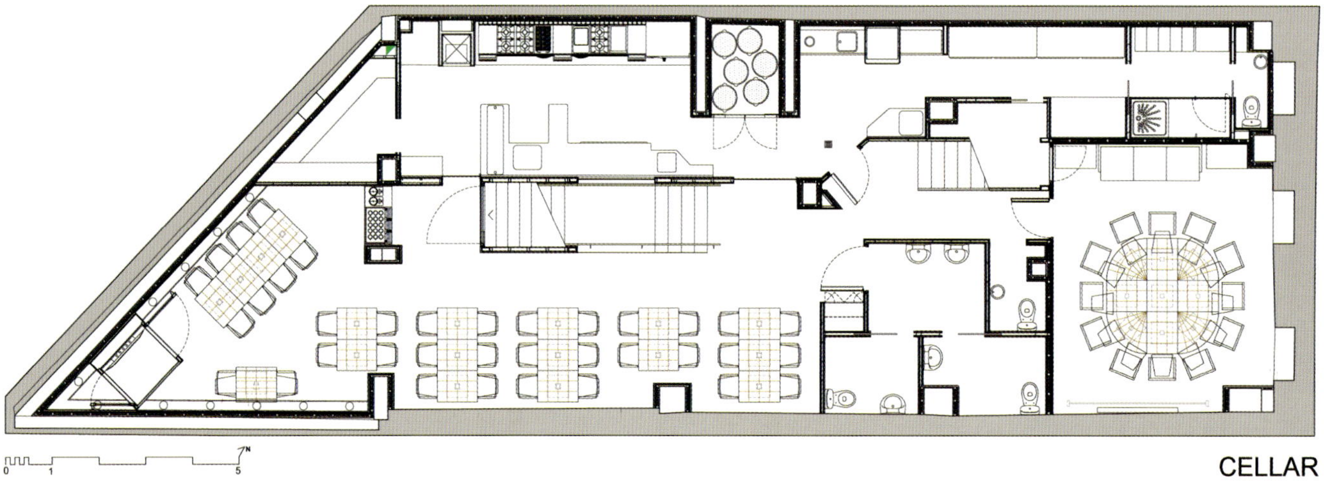

CELLAR

LA ESPiGA

BAR LA ESPiGA

Ground floor: Continuity of the white pavement of the sidewalk inside of the bar. The ceiling and back wall of the bar are covered with natural wood, with a stratified scheme reminding the guipuzcoan cliffs. The sculptured bar of oxidized metal and the retro-illuminated wine rack, accompany us from the street entrance to the interior of the bar.

地面一层：白色路面的人行道一直延伸到酒吧内。酒吧的天花板和后墙用原木覆盖，通过使用分层的设计形成悬崖峭壁的效果。从街区入口到酒吧内部，均设有氧化金属制成的带有刻纹的栅栏和壁照式酒架。

In the back, an allegorical mural of the Cantábrico, a painting by Edward Chillida Belzunce, extends down to the underground floor through a patio of natural light.

In the underground floor: The ambiance of the gastronomic society is recreated, a cultural and social aspect of the city. A stair of oxidized sheet composes the space. The kitchen is integrated to the restaurant. The bathrooms constitute a separation between a private lounge with a direct access from the street. The continuous floor is composed with the other characteristic paving of the city: the sand of the beach of "la Concha".

The quiet is a tribute to its loyal customers. This space is dominated by a mural painter Edward La Borde. It is the brief history of the "Bar La Espiga", since 1929.

后墙上挂着 Edward Chillida Belzunce 所绘的寓言壁画：《Cantábrico》。自然光线通过天井照射到地下一层。

地下一层的环境重建一种美食社交的氛围，反映城市的文化和社交等方面。厨房也被整合到餐厅中。盥洗室将私人会所同通往室外的通道分隔。地板的设计融入了城市道路铺设的特点，好似 la Concha 沙滩的感觉。

餐厅为客人提供了安静的就餐环境。墙上悬挂壁画大师 Edward La Borde 的作品，展示了自1929 年以来"La Espiga"吧的历史。

荷兰餐吧

Project > 项目名称 > 荷兰餐吧

Architect > 设计公司 > Creneau 国际

Location > 项目地点 > 荷兰阿姆斯特丹

Creneau International
ideas of the anti-ordinary

项目介绍 PROJECT DESCRIPTION

HMSHost Schiphol Airport, the leading hotel and catering provider at the airport, called on the atmosphere architects of Creneau International to design the Dutch Kitchen & Bar. The entrance immediately sets the tone, as you go under an arched hedge of oversized cutlery. The designers have also toyed with scale and proportion in the interior, where visitors can relax in a tea cup of Delft blue service, or try typically Dutch specialties at the "Mad Tea Party" table.

荷兰史基浦机场的酒店和餐饮提供商 HMSHost 集团要求 Creneau 国际根据环境设计一家荷兰餐吧。餐吧的入口处反映了设计的整体基调，客人走过拱形的篱笆就看到了超大的餐具模型。另外，设计师并没有严格地划分内部尺寸和比例，客人们可以在"代尔夫特蓝陶瓷杯"上惬意地享受，也可以尝试荷兰独有的特色"风帽子旋转杯"。

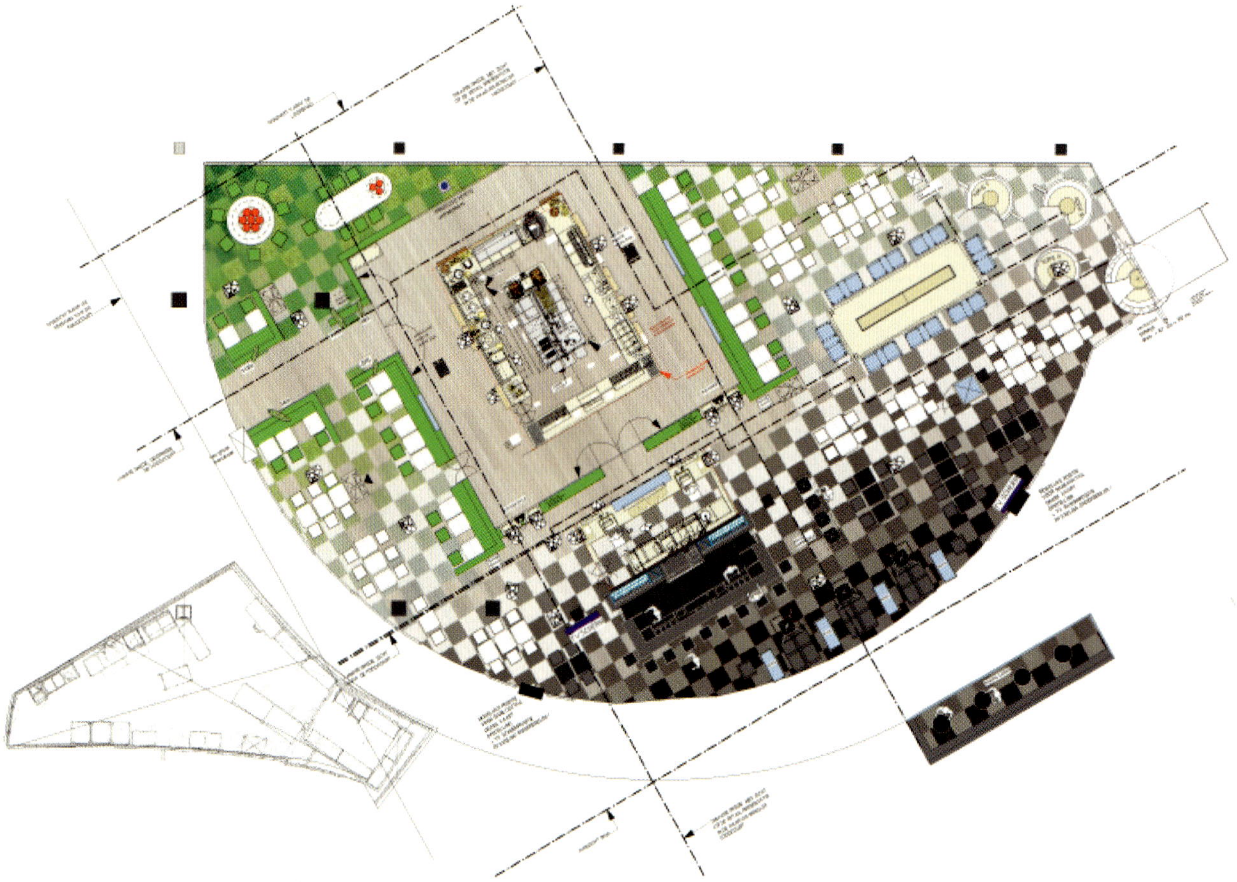

DUTCH BAR& COCKTAILS
BOLS COCKTAILS
Smoking area

The Netherlands may well be a very small country, but the Dutch are dreamers by nature and like to push back boundaries. Whether in technology, design, fashion or architecture, they often manage to outdo themselves. So the concept was self-evident: "where a small country can do great things." It had also the idea of creating a Dutch setting in surrealistic surroundings. Typical Dutch elements such as tulips are enlarged enormously out of proportion, while a view of a Dutch polder landscape on the floor from above is minimised tremendously. The original wall tiles and the service sets are painted by a Royal Delft master painter.

荷兰是一个不大的国家，但荷兰人天生是梦想家，他们喜欢以构思推动现实。无论在技术、设计、时尚或是农业方面，他们总是不断超越自己。这间餐吧的理念不言而喻：小国家能做大事业。设计同时融入了超现实主义元素。设计中包含了典型的荷兰元素，如被放大很多倍的郁金香装饰，从天花板到地板上都有微缩的荷兰景观画。原有的墙面和服务区都以皇家代尔夫特制陶工艺粉刷。

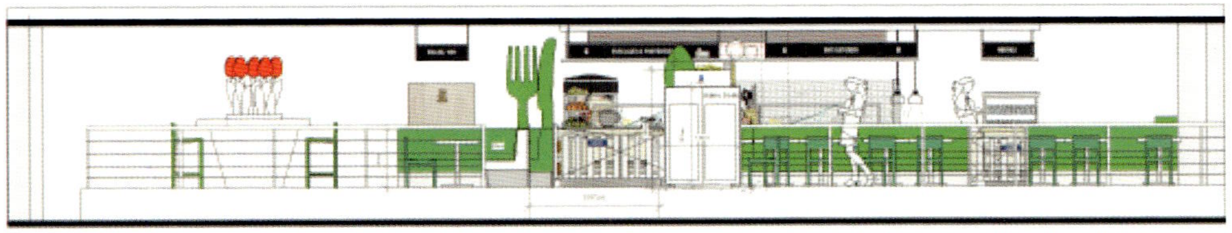

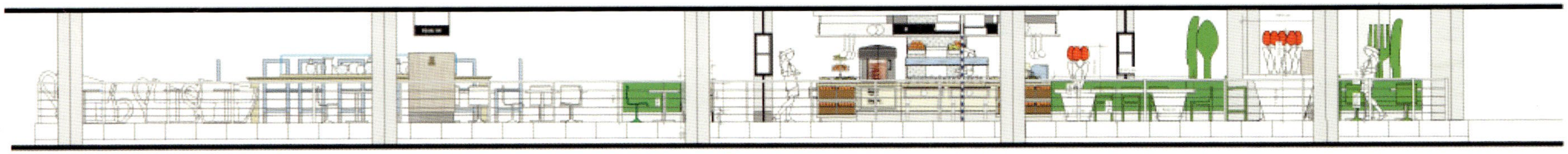

AANZICHT VANAF DE LOOPBAND

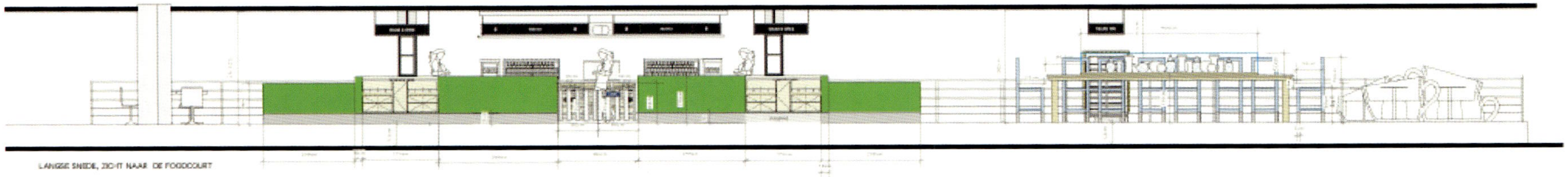

LANGSE SNEDE, ZICHT NAAR DE FOODCOURT

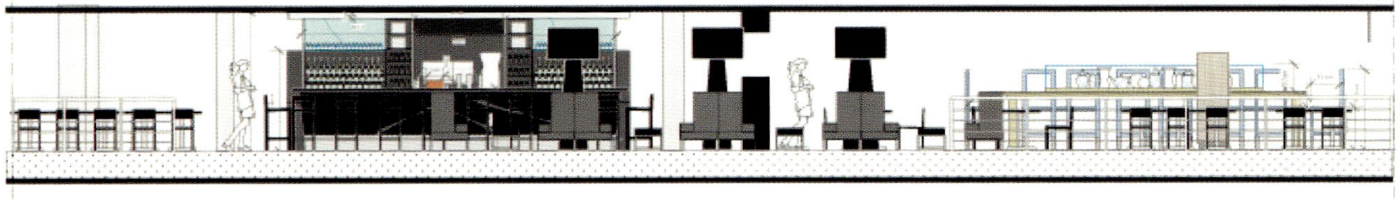

THANK YOU

The Dutch Kitchen and Dutch Bar are two different worlds that subtly merge thanks to a chequered floor pattern. The two worlds reflect the characteristic image of the Netherlands: classic on the one side, with houses on the canals and tulips; modern on the other side, with contemporary architecture.

荷兰餐吧的厨房和吧台是两种不同风格，如此精细自然的融合源自于多变的地板方格图案。这两种不同的风格形象地反映了荷兰的特点：一边是经典的传统风格，房屋建在河渠上，周围簇簇着郁金香；另一边是现代风格，建筑风格现代新潮。

SOUP & SALAD
CASHIER & RETAIL

PANCAKES POFFERTJES
SOUP & SALAD
SANDWICHES
FRESH JUICES

Interior architect Lieve Vandeweert: "We opted for expressive colours and materials to give visitors a real feeling of two worlds. The Dutch Kitchen exudes more charm, with light, bright colours, whereas the Dutch bar has a rather cosmopolitan and adult look, with mysterious, dark hues and gleaming textures."

室内设计师 Lieve Vandeweert 说："我们选取了具有表现力的色彩和材料，使参观者对两种不同风格有明显的对比。厨房是十分吸引眼球的亮色，而吧台则是更为流行、充满神秘色彩的暗色，而且使用了闪光的材质。"

Maialino 餐厅

Project > 项目名称 > Maialino 餐厅

Architect > 设计公司 > Rockwell Group 设计公司

Location > 项目地点 > 美国纽约

Built area > 项目面积 > 372 m^2

Photographer > 摄影师 > Ellen Silverman

项目介绍 PROJECT DESCRIPTION

Rockwell Group is designing a new neighborhood Italian restaurant, and the goal of the design is to capture the feeling and the warmth of a traditional local trattoria, but reinvented in this contemporary setting.

Just inside the entrance is a warm, neighborhood café meant to lure in either New Yorkers from down the block, or hotel guests. At seated or standing tables placed on a stone mosaic floor, guests will be served either espresso, breakfast, lunch or dinner. An apothecary-inspired wine room off the café allows the space to transform into a wine bar in the evenings. Backlit blackboards at the bar will announce featured wines and the daily changing specials.

Rockwell Group 设计公司新设计了这家意大利餐厅，设计延续了当地传统餐厅的感觉和风格，但是在陈设上有所创新。

入口处温馨的咖啡厅吸引了纽约人驻足，同时也迎来更多的酒店客人光临。马赛克石材的地板上设有座位和站立桌，客人们可以在这里享用咖啡，也可以在这里吃到早、中、晚三餐。受咖啡厅边的藏酒室启发，这里晚上被用作酒吧。吧台上的背光式黑板定时发布特色酒品和每日供应菜单。

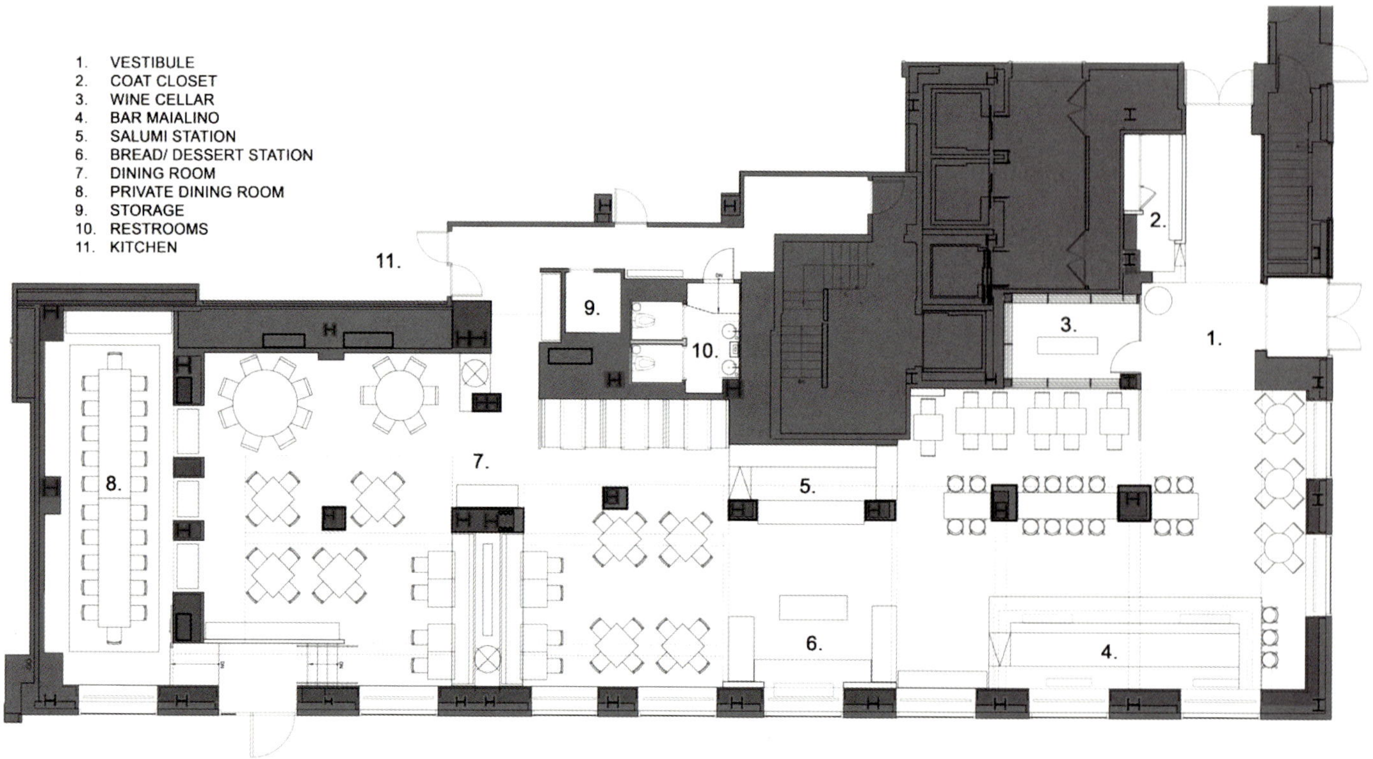

Cestino di Dolci-$11
Brioche Caramellato-$4
Cornetto-$3
Pane Cioccolato-$3
Budino di Olio d'Oliva-$3
Tortello di Agrume-$3
Pane e Marmellata-$4
Bomboloni con Crema-$4

No Italian trattoria would be complete without a traditional Salumi station. Between the café and the main dining is a stone-tiled area with two stations, one for meat and cheese, and another for desserts and bread. Both stations are custom made of stone, metal, glass and reclaimed wood. At the Salumeria-inspired butcher area, cheese and meat are plated for diners, with gourmet prosciutto and ham on display.

The main dining room is infused with the same elegant but casual feeling of an Italian trattoria, with windows overlooking the lush Gramercy Park. Loose tables and custom woven leather booths and banquettes are carefully placed throughout the space so that every seat would be a prime one, without ever losing the sense of informality. The floors are wood parquet, and the walls are covered with textured plaster and reclaimed wood wainscot. The space is anchored by a ceiling-height wine display wall, which also acts as a screen between the main and private dining rooms.

所有的意大利餐厅都有传统的腌肉操作台。咖啡厅和主餐厅之间是两个石质操作台，一个用于制作肉和奶酪，另一个用于制作甜点和面包。操作台的材料均为石头、金属、玻璃和再生木材。厨师在制作肉的操作台上将切好的肉装在盘子里递给客人，台子上还有用于展示的意大利熏火腿。

主餐厅的格调高雅，同时又兼具意大利餐厅的休闲特点，透过窗子可以欣赏到纪念公园的优美景色。宽松摆放着的餐桌、订制的编织皮质展台和走道无一不展示着空间布置的巧妙和用心，每个座位都经过精心的设计，但同时又不会过于正式和拘谨。地板采用木质材料，墙面涂上带有编织纹的灰泥，再生木板做成的壁板成为分隔主餐厅和包间的屏风。

The private area is filled with warm, saturated colors that create a rustic and home-style feeling for the large communal wood table that can seat up to 22 people. An oversize chandelier hangs over the table, and the floating wine display wall provides a filtered view into the main dining room.

包间的布置温馨，高饱和度色调营造了淳朴的家庭氛围，大木桌可以容纳 22 人就餐。大吊灯挂在桌子上方，人们透过用于展示各种酒的墙壁可看到主餐厅。

卡尔斯厨房

Project > 项目名称 > 卡尔斯厨房

Architect > 设计公司 > Concrete Architectural Associates

Location > 项目地点 > 荷兰阿姆斯特丹

Built area > 项目面积 > 1000 m^2

Photographer > 摄影师 > Ewout Huibers

concrete

项目介绍 PROJECT DESCRIPTION

Karls Kitchen is the new restaurant of luxury department store Breuninger in Stuttgart. The kitchen is the free-flow restaurant's central element. The entrance to the restaurant is characterized by large glass fridges with fresh products that also function as a display window.

卡尔斯厨房是斯图加特市 Breuninger 购物广场附近新开的一家餐厅。餐厅的风格包含了自由不受拘束的元素。餐厅入口处放有盛放生鲜食品的大型玻璃冰柜，同时也起到了展示作用。

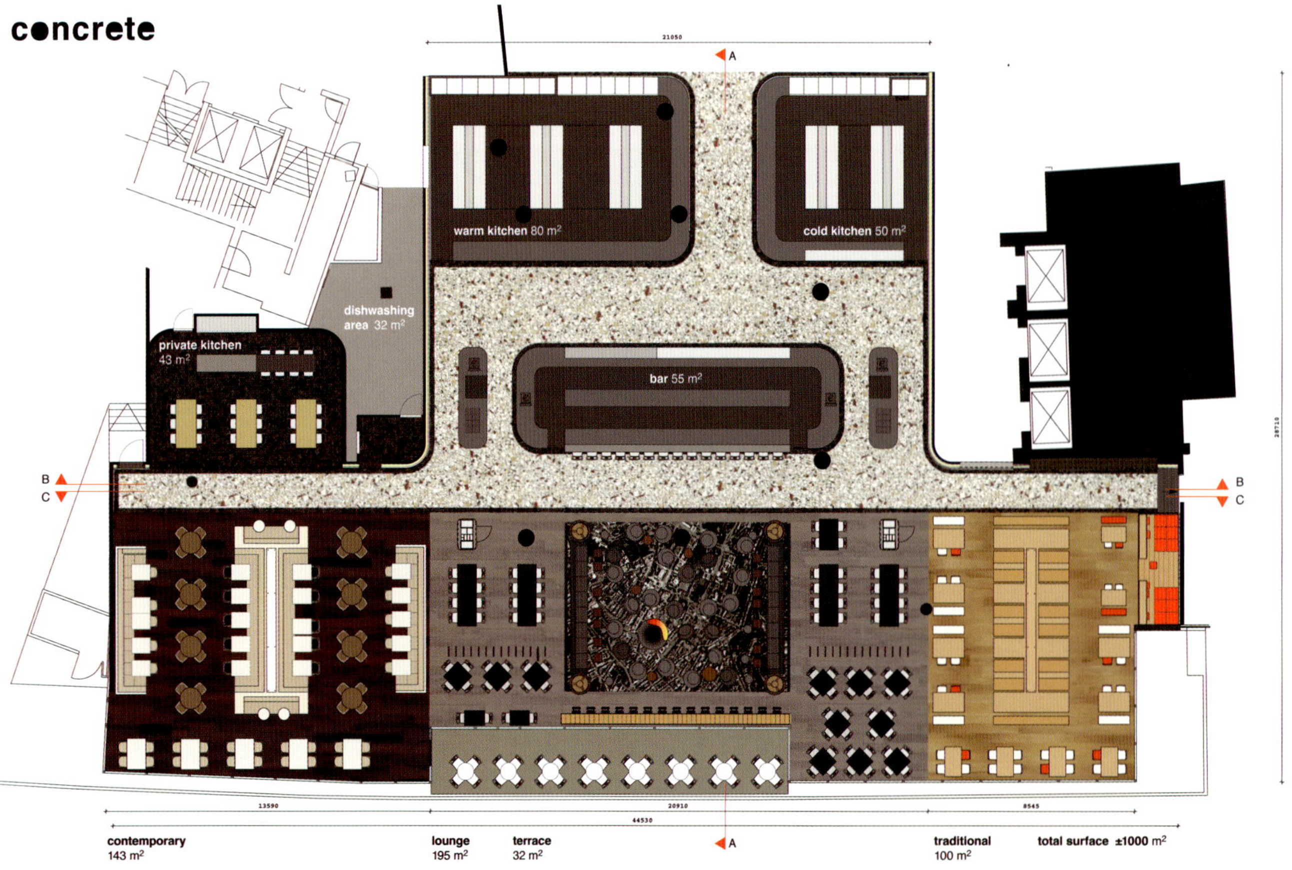

(karls) kitchen

100

101

The materials used reference the classic kitchen of grandma's time; white tiles, black natural stone in combination with a terrazzo floor. The modern design in combination with new materials such as glass and stainless steel create a modern design with a link to the classic kitchen. Opposite the kitchens is the bar, a free-standing element between the free-flow restaurant and the seating area. The bar is made of wooden tiles in the same dimensions as the kitchen tiles, on top of which is a black natural stone surface, creating a warm appearance and being functional at the same time.

厨房的装修参考了经典风格，白色的瓷砖、黑色的天然石板与水磨石地板相结合。同时，富有现代感的设计又加入了新材料的运用，如玻璃和不锈钢。两种风格有机统一，完美融合。吧台正对厨房，与就餐区相对独立。此处的功能具有双重性，既能做蛋糕等甜点，又提供新鲜果汁和其他酒水。构成吧台的木板与厨房的瓷砖尺寸相同，表面是黑色天然石板，营造了亲切的氛围，同时兼具实用性。

(karls)
kitchen

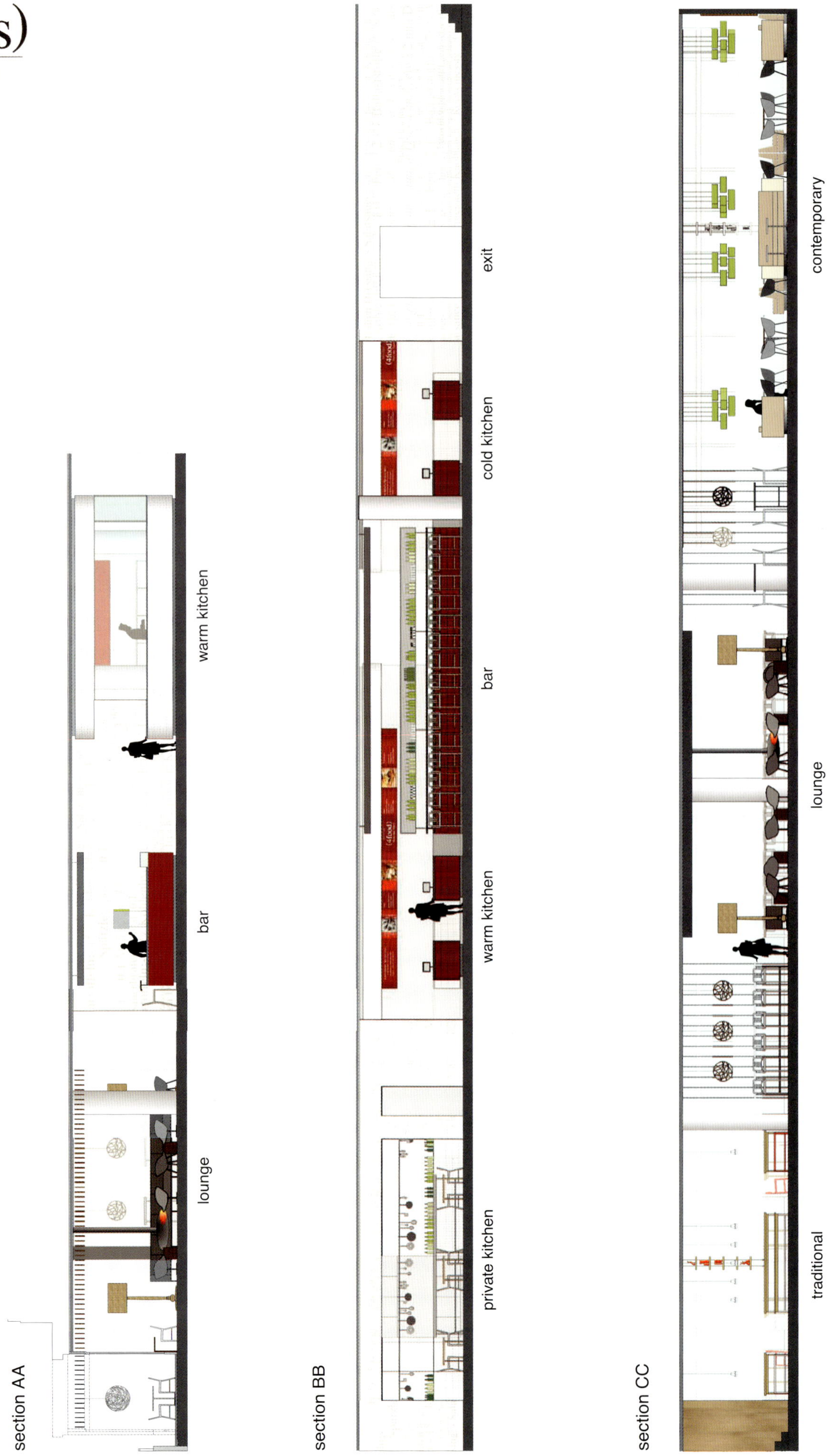

The restaurant is divided into 3 different seating areas, each with its own identity and character. First of all there is the traditional seating area, which refers to Breuninger's history and the traditions within the department store, as well as in Stuttgart. The modern seating area represents the Stuttgart of today, and the Lounge is the refuge and, at the same time, the connection between the modern and the traditional seating areas.

餐厅分为三个就餐区，每个区域都有不同的特点。第一个是传统风格的区域，与斯图加特和 Breuninger 购物广场的历史相符。富有现代气息的座位区代表了斯图加特的今天，休息室是现代区和传统区的连接点。

The modern seating area is defined by contemporary materials, such as Corian, light-coloured leather, chrome in combination with dark wood. In this cabinet there are cooking utensils in silver and chrome which reflect towards the modern character. The dark wooden floor creates a subtle contrast with the light furniture. The custom-made hanging lamps, with the different dimensions and shades of green, provide a playful touch. Alongside the three seating areas there is the restaurant's private kitchen, which can be closed off by means of a sliding glass door. The room has its own free-standing cooker and can be used for cooking workshops, parties, presentations or business meetings.

富有现代气息的座位区使用了当代流行的材料，如人造石、彩色皮革、铬以及黑木。中间的橱柜放着具有现代特点的厨具。暗色调的地板与亮色的家居形成鲜明的对比，私房菜厨房位于三个座位区的旁边，用玻璃推拉门分隔。这里有三个独立厨灶，可用于研讨会、聚会、发布会或是为商业宴会准备。

“工厂” 餐厅

Project > 项目名称 > “工厂”餐厅

Architect > 设计公司 > Tjep. 设计公司

Location > 项目地点 > 荷兰

Photographer > 摄影师 > Daniel Nicolas

项目介绍 PROJECT DESCRIPTION

The clients commissioned a new restaurant to be located in the Rotterdam harbor, right between the boats and the cranes. Fabbrica struck us as the most convenient name for this new Italian restaurant. Fabbrica meaning factory in Italian, designers envisioned the canteen of a very special factory: a very romantic factory where pleasure is produced for guests.

客户委托设计师在船与起重机繁多的鹿特丹港设计一家新餐厅。对于这家新意大利餐厅，设计师的第一感觉是Fabbrica这个名字再合适不过了。Fabbrica在意大利语中意为"工厂"，设计师想打造一家非常特别的"工厂"餐厅，一个非常浪漫的，可以为客人带来快乐的"工厂"。

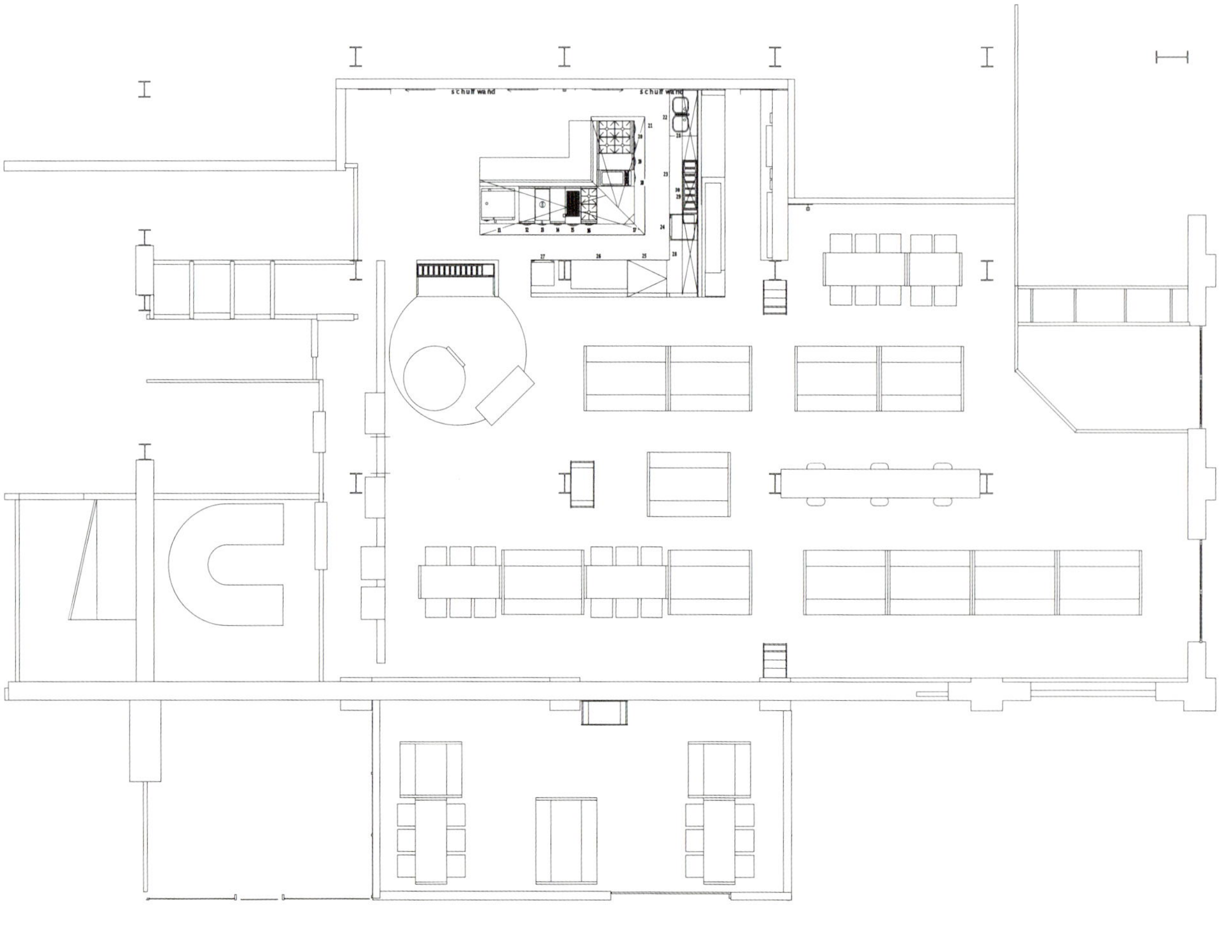

Designers chose not to intervene with the authentic industrial character of this nineteenth century warehouse so they left all the structures in original state. The walls for example are left intact, in several places large glass panels were placed in front of them covered with Italian wallpaper patterns. The result is that the patterns seem to float in front of the wall in a complex game of reflections. Lovers get to sit in a train structure that floats in the center of the space. The logo of Fabbrica is based on a font in which the point on the i resembles the beautiful shape of a hand made pizza, but one can also see a full moon, as it enlightens Fabbrica at night.

设计师不想破坏这间 19 世纪仓库的工业特性，所以所有的构造都保持原状。例如，墙壁原封不动地保留了下来，在好几处墙壁前面放置了覆以意大利墙纸图案的大型玻璃嵌板，可利用影子营造出一种图案在墙壁前面漂浮的感觉。恋人可以坐在大厅中央的车厢中。Fabbrica 标志所选用的字体中字母 i 上的一点类似意大利手工披萨的形状，也可将其看成一轮满月，会在夜间将 Fabbrica 照亮。

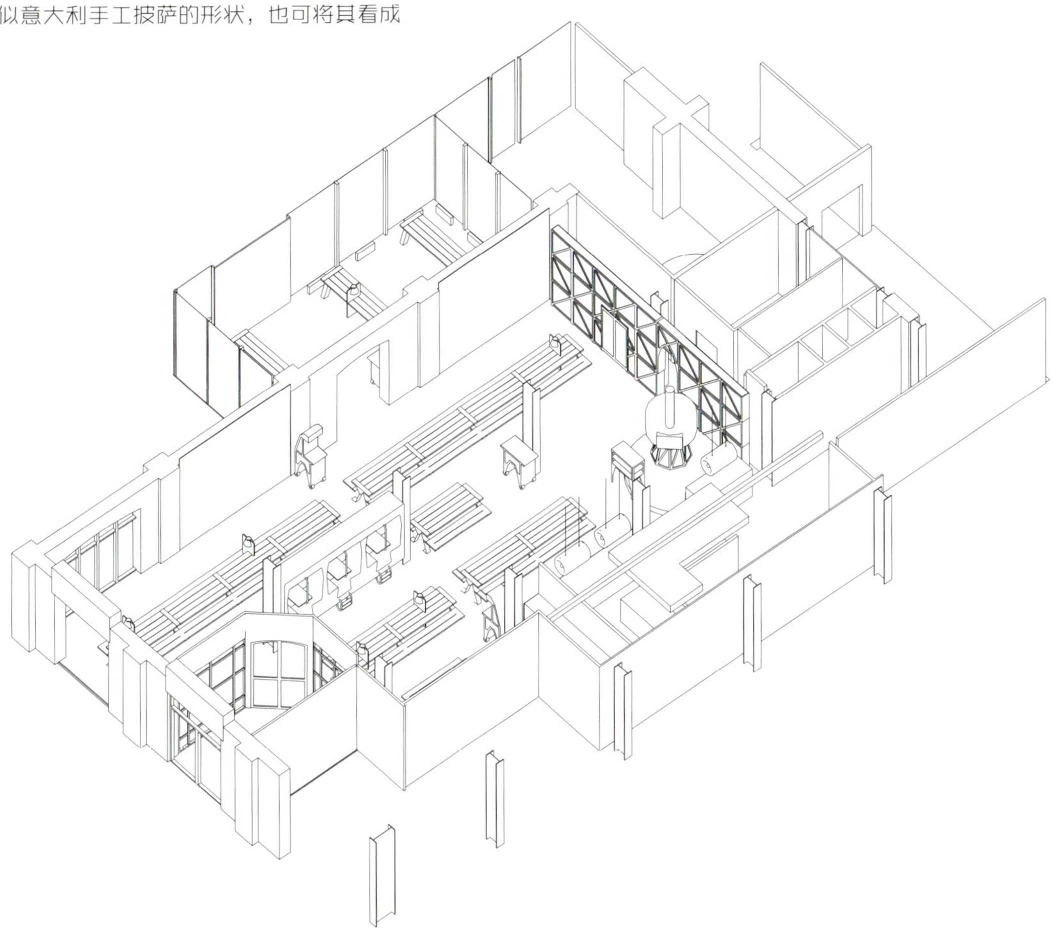

Like in the canteen of a factory you will find long tables and benches at Fabbrica, but than colored in Italian ice-cream shop style colors: pink and pistachio green. The oven is placed in a huge tank, covered with Italian mosaic. A large wall composed of crane elements is used to store wood to fire the oven. Every detail combines industrial sturdiness with soft colors or decorative elements.

如同走进工厂的餐厅，在 Fabbrica 你会发现许多长桌和长凳，但颜色采用意大利冰激凌店的色调——粉色和浅果绿色。炉子放在一个巨大的水箱里，上面砌以意大利式马赛克。一面由起重机元件构成的墙用来储藏炉子燃烧所需的木头。每一处细节既有工业的粗犷，又施以柔和的色调或装饰性元素。

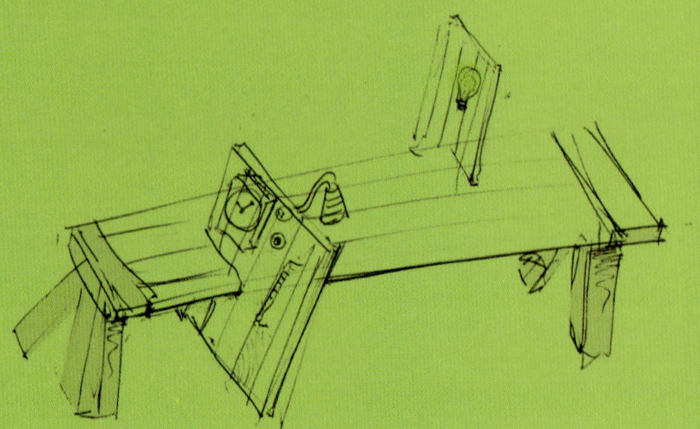

fabbrica

梅尔卡餐厅

Project > 项目名称 > 梅尔卡餐厅

Architect > 设计公司 > VAíLLO & IRIGARAY + GALAR 设计公司

Designer > 设计师 > Antonio Vaíllo i Daniel、Juan L. Irigaray Huarte 、Daniel Galar Irurre

Built area > 项目面积 > 990 m^2

Photographer > 摄影师 > José Manuel Cutillas

m'o

RESTAURANTE ELMERCAO

项目介绍 PROJECT DESCRIPTION

We like the restaurants that change: that every time you go are different, changing your emotions with the type of event you want, with the environment you want to recreate. We hate the feeders. We like places that speak of where you are, the climate in which they are involved, what they have at them side. We like the restaurants that have appropriate corners, dwellings which can be tailored to each. We seldom have found, perhaps that is why we accept the challenge of build a restaurant in Merca.

我们喜欢富有变化的餐馆：每次的到来都有不同的感受，心情可以随着周围的环境变化，同时也可以放松心情。我们讨厌为吃饭而吃饭。我们喜欢有个性的地方，喜欢融入其中的氛围。我们喜欢设置有适当的角落和房间的餐厅，可以供设计专门发挥……但是这种地方很难找到，这就是我们接受梅尔卡餐厅设计的挑战的原因。

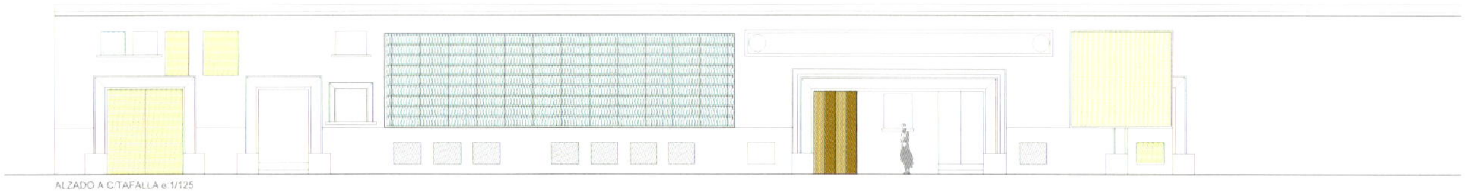

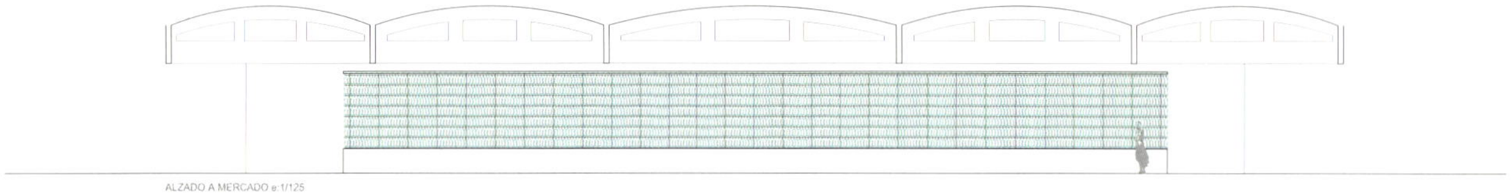

The restaurant occupies two floors: the first connected with the restaurant and the second located in semi-basement. It also offers two spaces, two different modes of being: a daily, open, flexible, linking to the circulation of the street, organizing the circulation and access through the bar which has deeper, austere atmosphere with the concept of wine; the place underground providing a receptacle restful, serene and peaceful.

餐厅分为两层，一层直通大街，一层是地下室。在内部有两种空间，两种氛围：一种是大众的、开放的、灵活的，同大街相连，重点是行径线的设计和酒元素的运用；而深入到地下的酒吧，环境更为安静、祥和。

el mercáo
RESTAURANTE

The designer essentially aims to create an atmosphere: warm, intimate, understated, timeless, moody, elegant oriental, where the light can indiscriminately mix the day and night, party and privacy.

设计师想通过设计达到这样一种空间氛围：温馨、亲切、低调、永不过时、带有感情色彩并具有东方神秘而优雅的特点，这里的灯光能巧妙地融合昼夜、兼顾聚会与营造私密感。

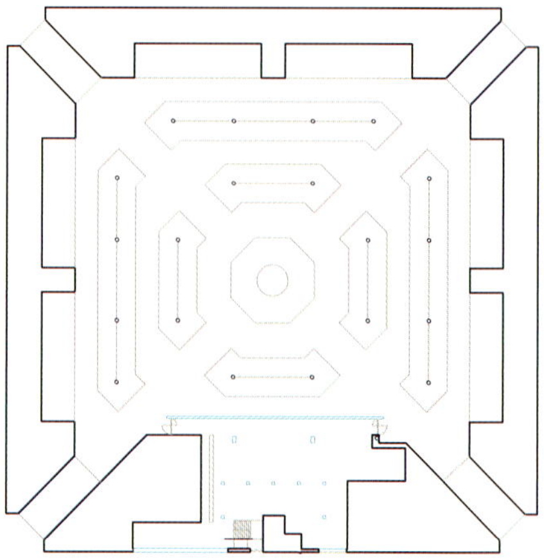

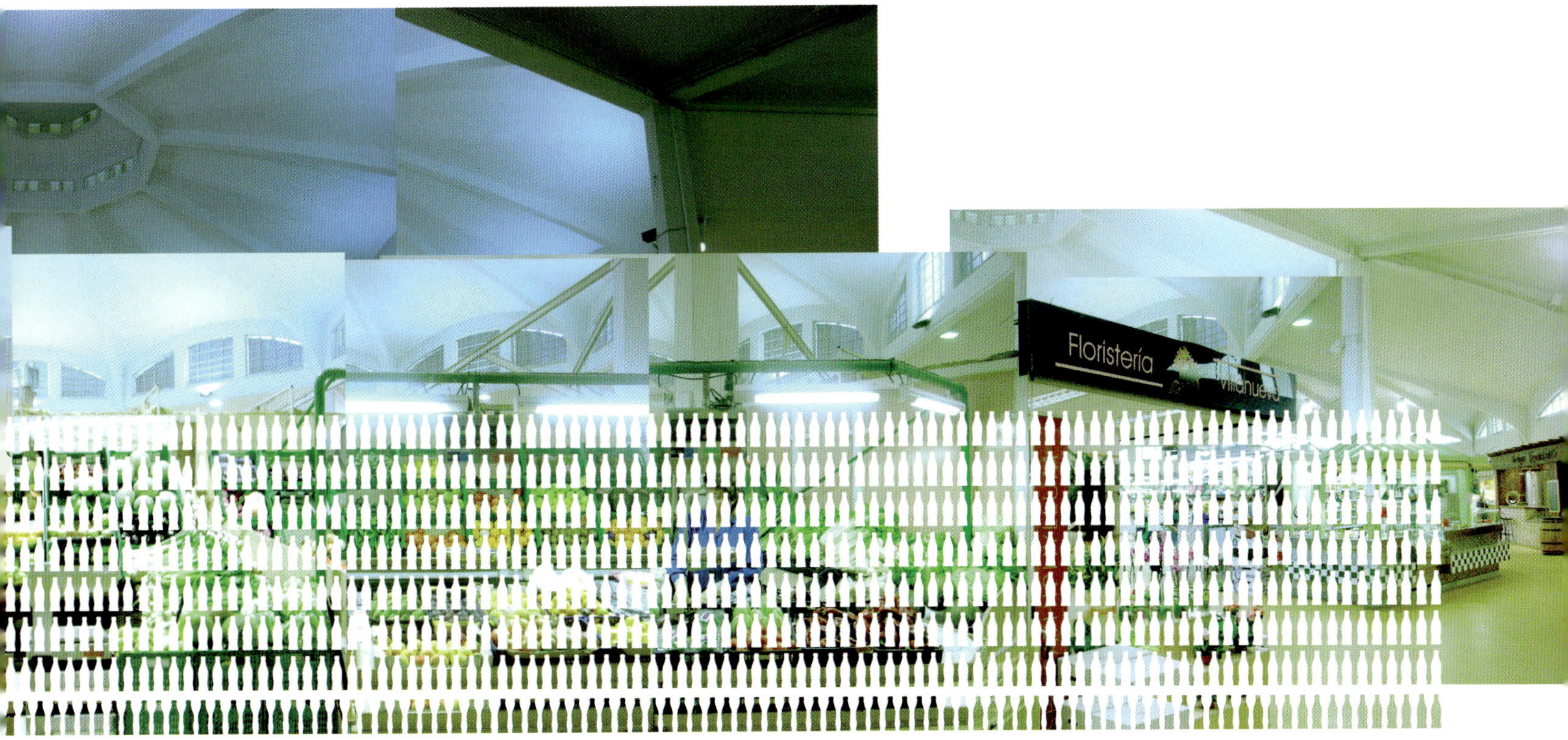
Floristería
Villanueva

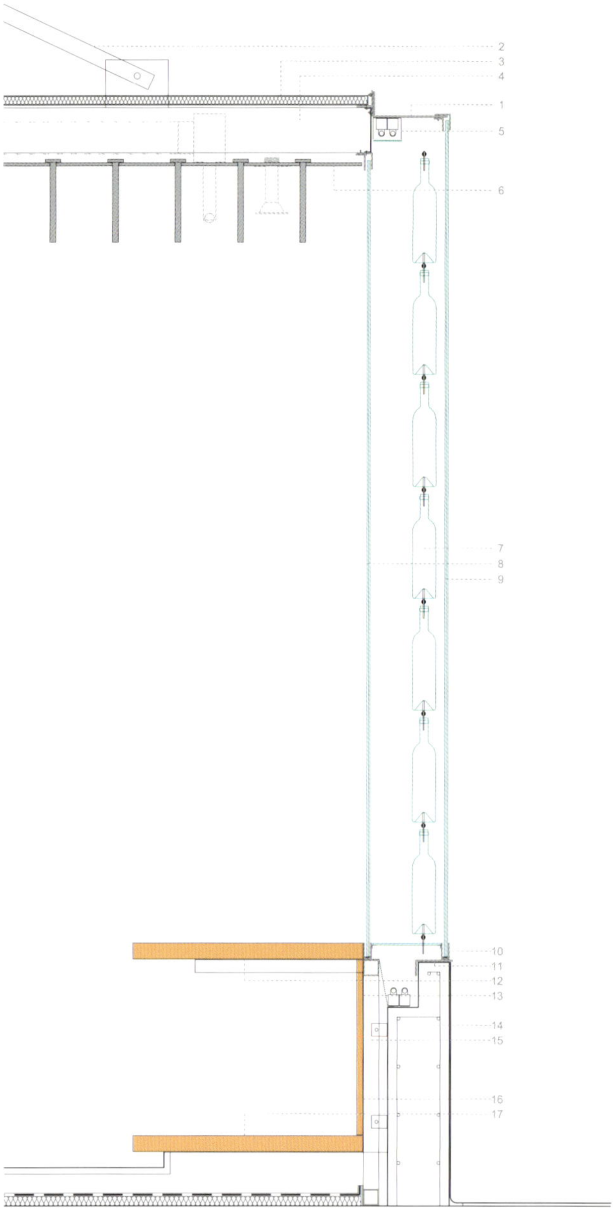

SECCIÓN BOTELLERO A MERCADO

DETALLE ALZADO BOTELLERO A MERCADO

1 PERFIL GALVA. PINTADO L230.6.8
2 TIRANTE: PLETINA GALVA. PINTADO 50.15 UNO POR CADA PER
3 SANWICH CHAPA MICROPEFRILADA ASILAMIENTO LANA ROCA
4 SUBESTRUCTURA CF140.2.0 C/120 CM
5 FLUORESCENTE CON DIFUSOR DE PLEXIGLAS SATINICE VERD
6 REGISTRO ILUMINACION
7 BOTELLA BORDELESA 057/242 DE VIDRALA COLOR OSCURO
8 VIDRIO INTERIOR 5+5
9 VIDRIO EXTERIOR 5+5
10 CARPINTERIA GALVA. PINTADO U30.65.4
11 PERFIL APOYO L110.50.5
12 MESA DE MADERA DE IROKO ENLISTONADO e:3cm
13 REGISTRO ILUMINACION
14 MURO DE H.A. + PINTURA EPOXI
15 SUBESTRUCTURA: TODO #50.3
16 TAPA DE MADERA DE IROKO e:2cm
17 SUELO DE MADERA DE IROKO e:5cm

Spaces are split by curtains that reflects to the light. It's made of velvet (good touch and optimum acoustic performance), acting like large veils that in the orthogonal direction with ceiling. The almost theatrical scenery makes no partitions, no concept of closed box, it generates an ambiguity border among a series of veiled spaces. Its connection with the outside world is generated through the filters: glass aquarium filled with green glass bottles that filters light and views.

空间用天鹅绒窗帘隔开，能够透光，既有很好的质感，又能隔音。这些窗帘在天花板上的装饰间形成直角，如同许多巨大的幕布，又不会让空间显得封闭。室内通过外立面玻璃缸里的绿玻璃瓶过滤光线和景物，与外界相通。

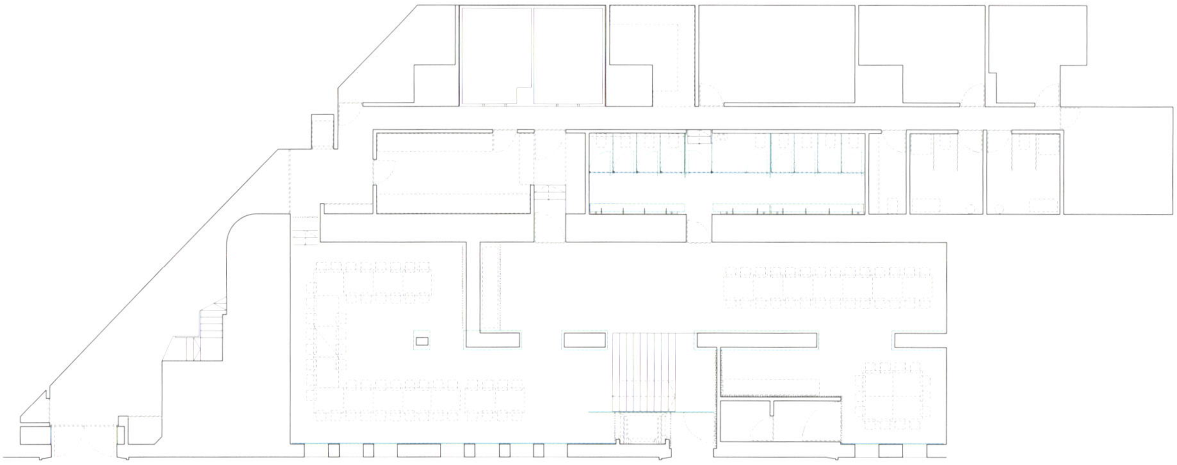

PLANTA SEMISÓTANO e:1/200

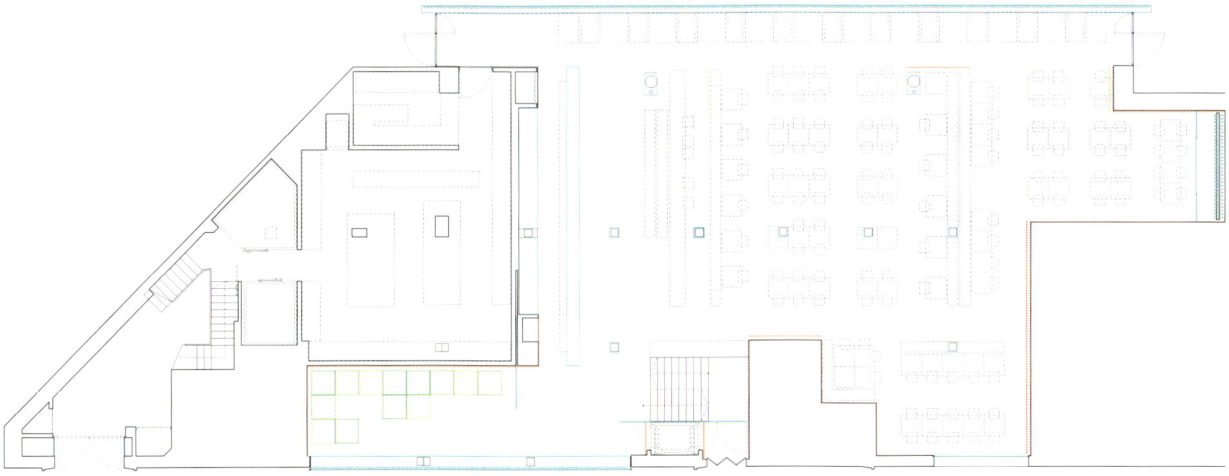

PLANTA BAJA e:1/200

The corrugated roof with a grooved surface, dark accents and gestures accompanied by undulating spaces that organize a geometric replica on the ground, wet bar, high table, bench, there are a variety of tables reinforce the different ways of being: low, high, large and small, square and oblong, and so does the mode of sitting: chairs, benches , banks high or low, individually shared up the different ways of eating. The objects placed over carpets (wood benches and tables) unify the color and texture appling massive slatted iroko boards.

带横楞的天花上呈暗色调，与整个空间的风格相搭配，地板、吧台、高脚桌、长沙发都有相同的几何感。室内有许多高矮、大小、形状不同的桌子可供客人选择，同时配有各式各样的座椅：靠椅、长凳、斜椅，让客人有不同的进餐体验。摆放在地毯上的长凳和桌子用巨大的绿柄桑木板制成，色彩和质感统一。

Dishroom 餐厅

Project > 项目名称 > Dishroom 餐厅

Architect > 设计公司 > Afroditi Krassa 设计公司

Location > 项目地点 > 印度

Built area > 项目面积 > 465 m^2

Afroditi krassa

项目介绍 PROJECT DESCRIPTION

In 2008, three entrepreneurs venturing into the restaurant world approached Afroditi. The brief was simple: design us a contemporary Indian restaurant. Afroditi's approach was clear from the beginning: no stereotypes, no clichés, let's design something totally unexpected. Some months of research and a few trips to Bombay later, the Dishoom identity was born.

2008 年，进军餐饮业的三大企业家找到了设计师 Afroditi。他们的要求很简单：设计一家当代印度餐厅。从一开始，Afroditi 的定位就很清晰：摒弃陈规陋习，设计一家前所未有的餐厅。于是，经过几个月的研究，以及随后的几趟孟买之行，Dishroom 餐厅诞生了。

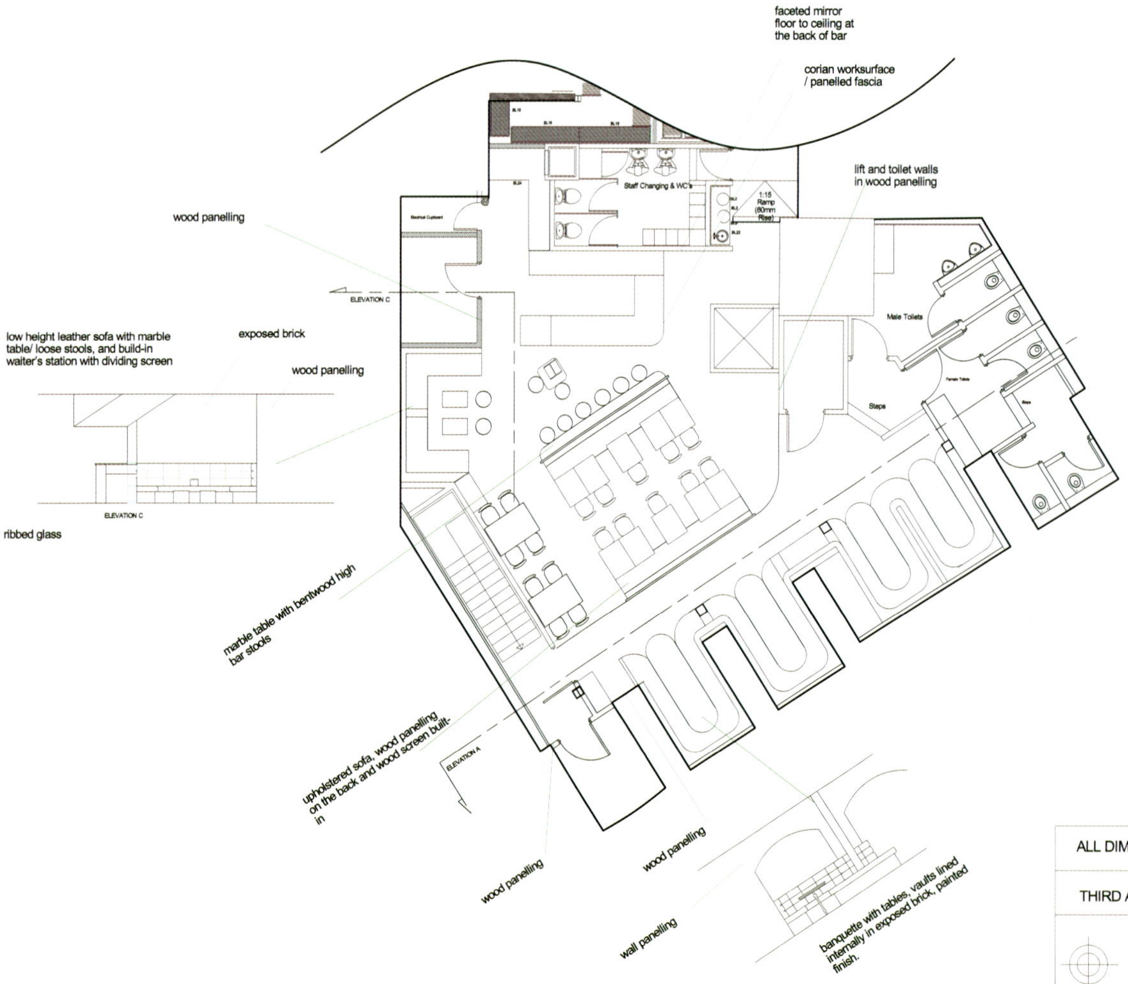

ALL DIMS IN MM	Afroditi krassa	
THIRD ANGLE	BASEMENT FLOORPLAN	
	DRAWING NO A/1016	SCALE 1:100
	DATE: 14.12.2009	DISHOOM

RULES OF THE CAFÉ
NO SMOKING
NO FIGHTING
NO CREDIT
NO FOOD FROM OUTSIDE
NO TALKING LOUD
NO SPITTING
NO BARGAINING
NO CHEATING
NO WATER TO OUTSIDERS
NO MATCHES
NO GAMBLING
NO COMBING HAIR
ALL CASTES WELCOME
POND'S
SABITADEVI in
Dr. MADHURIKA
or MODERN WIFE.
with:- MOTILAL
डॉ. मधुरीका
STORY BY:- Sjt. K.M. MUNSHI
DIRECTED BY:- SARVOTTAM BADAMI
"What does he do at school, sit and dream?"
HORLICKS

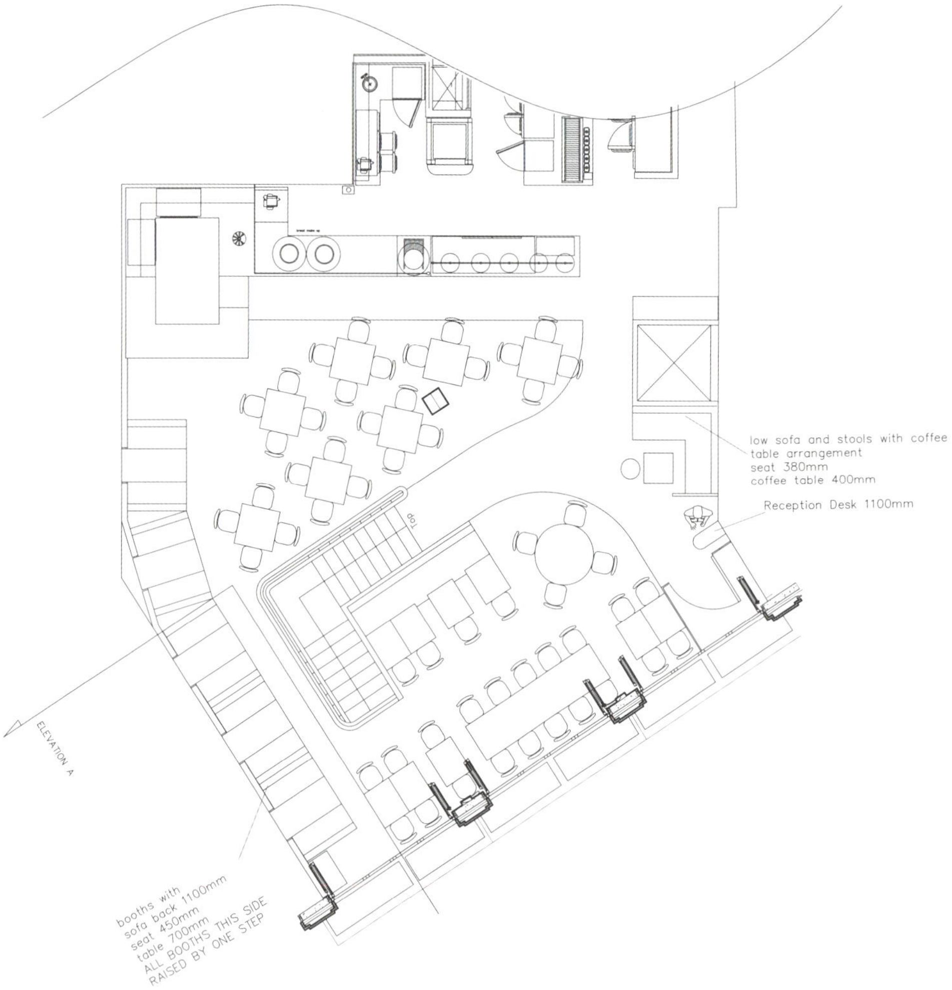
low sofa and stools with coffee table arrangement
seat 380mm
coffee table 400mm
Reception Desk 1100mm
ELEVATION A
booths with
sofa back 1100mm
seat 450mm
table 700mm
ALL BOOTHS THIS SIDE
RAISED BY ONE STEP

Located in Upper St Martin's Lane and occupying a site of 5,000 square feet, Dishoom is an all-day Bombay café. "Dishoom", by the way, is an old Bollywood term, an expression a little like 'Pow!' or 'Bam!' which also encapsulates the optimistic energy of the concept and design.

Dishroom 餐厅地处上圣马丁路，占地面积 465 平方米，全天侯营业。另外，"Dishroom"是宝莱坞一个古老的词语，意思近似"啪"或"叭"，这也包含了乐观积极的设计理念。

A strict, monochromatic colour palette has been used throughout; in combination with retro Bombay portraiture and 60s graphical, pop imagery. Subtly exotic and classically urban, Dishoom is a far cry from UK Indian restaurants, avoiding all expected clichés that come hand in hand with the territory, yet respecting the overlooked beauty of everyday India.

餐厅上下除了严格采用一种单一色调之外，还采用了一些复古的孟买肖像画、20 世纪 60 年代画作及流行画作。既古典又现代，同时又颇具异国特色的情调，使 Dishroom 远不同于英国一般的印度餐厅：既摒弃了所有预料中的人所熟知的印度元素，同时又保留了印度日常生活中被人忽视的美的一面。

Rather than looking into some of the blunt, contemporary architectural offerings of Bombay, the studio drew from the city's rich past and the honest, resourceful and functional interiors that have developed throughout its chequered history. Old and fading Irani cafés, street stalls and other down-to-earth eateries, all perfect examples of an unexpected and classic Bombay charm, provided the design direction. The logotype pays respect to classic fonts, old school vignettes and the visual appeal of Hindi. The interiors combine elegant, sophisticated and simple lines with an influence from Bombay's art deco past. The checkerboard tiles, oak panelling, white Carrara marble topped tables and mismatched chairs create a relaxed, democratic café style space, in keeping with the food offering.

工作室没有参考孟买一些毫无特色的当代建筑作品，而是从这座城市历史中真实、丰富而实用的内在中汲取营养。古老而日渐没落的伊朗咖啡馆、露天商店及其他随处可见的小餐馆……所有意想不到而又经典的可以体现孟买风情的绝佳案例都会为设计提供导向。标志采用经典字体、旧式插图及印地语制造视觉吸引。室内设计受孟买过去的装饰艺术影响，糅合典雅、精致而简洁的线条。棋盘瓷砖、橡木镶板、白色镶以卡拉拉大理石的桌子以及错落布置的椅子在供应美食的同时，又营造出一种舒适而大众化的咖啡馆式氛围。

TWIN

龙舌兰酒餐厅

Project > 项目名称 > 龙舌兰酒餐厅
Architect > 设计公司 > CCS 设计公司
Location > 项目地点 > 美国
Built area > 项目面积 > 604 m^2
Photographer > 摄影师 > Melissa Werner

项目介绍 PROJECT DESCRIPTION

Reposado brings upscale Mexican dining to Palo Alto. Owner Rob Fisher and Head Chef Arnulfo Hernandez serve an inventive interpretation of authentic Mexican dishes. CCS designed this restaurant in an existing 1920's feed & grain warehouse, which previously housed a Café. Wit h 6,500 square feet, the restaurant seats 180 between the main dining space and large mezzanine.

龙舌兰酒餐厅为帕洛阿尔托市一家高档的墨西哥餐厅。店主与主厨对传统的墨西哥菜肴进行了大胆的创新演绎。餐厅原为 20 世纪 20 年代建造的饲料谷物仓库，在此之前曾是一间酒吧，后经 CCS 设计改造成为眼下的餐厅。餐厅面积为 604 平方米，在主用餐区及大的夹层间可容纳 180 个座位。

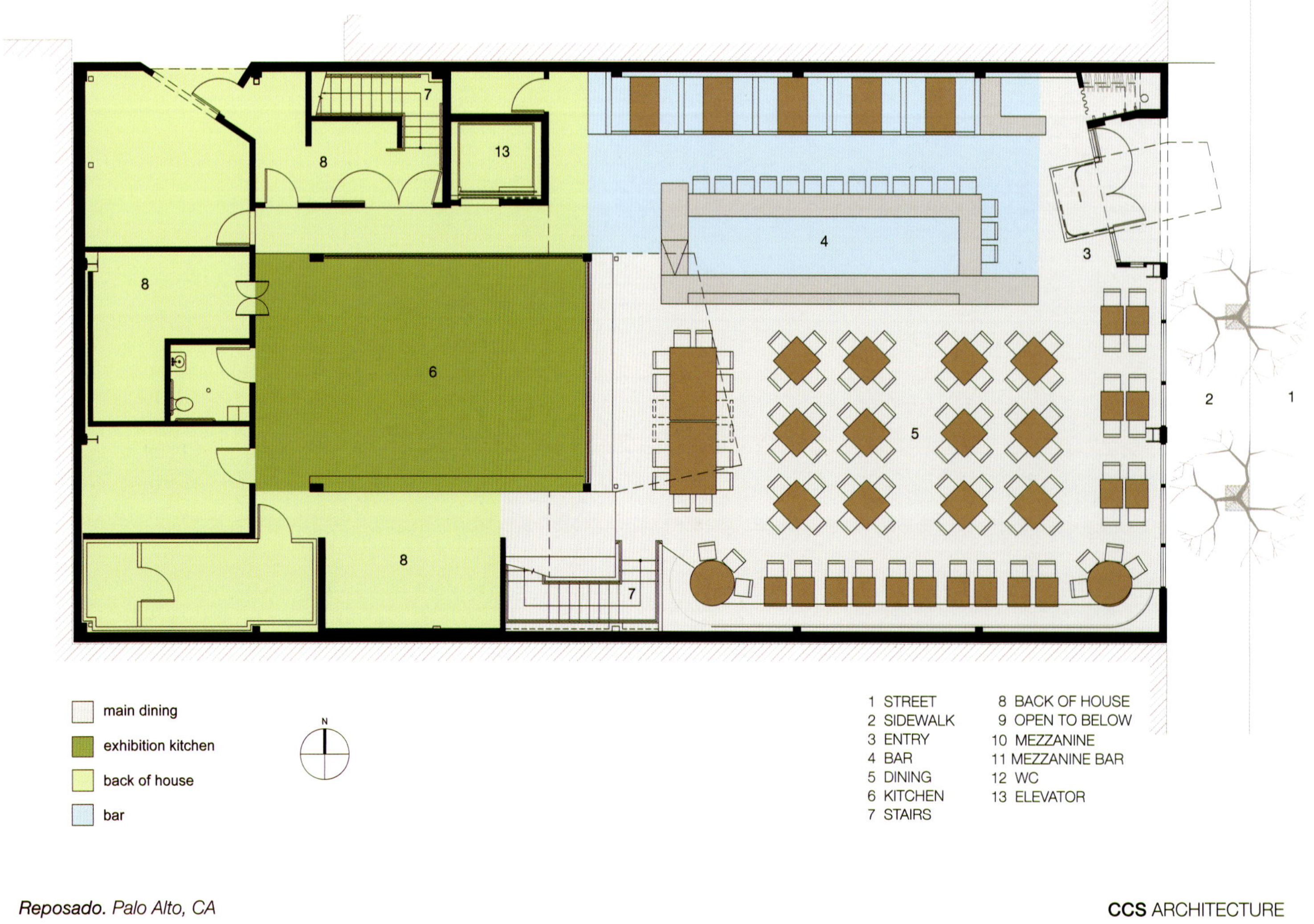

Located on Hamilton Avenue in the heart of downtown Palo Alto, the buildings historic façade has been modernized and painted a monochromatic blue, with the exception of the storefront windows. Mahogany wood entry doors, stainless steel projected canopy, and large signage lettering, accentuate the entry. A series of large sliding windows open up to allow for an uninterrupted outdoor connection to the main dining.

餐厅位于帕洛阿尔托商业区中心的汉密尔顿大道上。建筑物历史悠久的外围被重新改造，除临街窗户外，全部粉刷成了清一色的蓝色。红木大门、不锈钢顶蓬以及大型的雕牌刻字将入口衬托得气势恢宏。一系列大型推拉窗使室内更加明亮，同时又不会阻挡主餐厅与外界的联系。

Upon entering, guests are met with a soaring existing timber ceiling and trusses, existing concrete walls, and an entirely new modern interior. The upstairs mezzanine is cantilevered dramatically over the warehouse style windows, which display the exhibition kitchen below. Colorful walls and fabrics juxtapose the raw concrete and timber. This is most apparent in the large back lit culturally inspired artwork which adorns the walls.

进入餐厅以后，映入眼帘的是高大而古老的木质天花板、屋架、古老的混凝土墙壁，以及崭新而现代的室内设计。楼上的夹层戏剧性地悬垂于仓库式顶部，使下面的展览厨房一览无余。彩色的墙壁及织品与粗制混凝土和木质并置，给人以强烈感官刺激。这点从装饰在墙壁上的大型背光的民族风格艺术品上最能显现。

Custom hanging color fabric pendants carry out the theme. Polished concrete floors, stripped color carpet, dark flamed stone, mahogany wood, and stainless steel round out the material palette which is upscale yet playful.

约定俗成的彩色悬挂织物诠释出主题。随处可见的抛光混凝土地板、带状彩色地毯、黑色火焰石、红木及不锈钢，不仅高档，而且充满趣味。

The restaurant includes some distinct dining areas. The double height space on the ground floor consists composes the main dining, with banquet seating along the wall and a large raised communal table adjacent to the kitchen. The main dining is separated by a 30′ long, stainless steel and glass back bar with raised bar dining booths against the wall. The cantilevered portion of the mezzanine upstairs comprises overflow dining and private event space. With its own bar, operable skylights and views to the downstairs, it is a truly unique dining experience.

餐厅的用餐区划分很明确。底层双倍高度的空间为主餐厅，宴会席位沿墙而设，厨房附近设有一张高大的公共餐桌。主餐厅由一个 30 米长的不锈钢及玻璃质地吧台隔开，吧台靠墙有突出的餐饮展台。楼上夹层的悬垂部分包含公共餐饮区及包间两种。餐厅设有自己的酒吧、可调式天窗，还能让顾客看到楼下的美景，实为独一无二的餐饮体验。

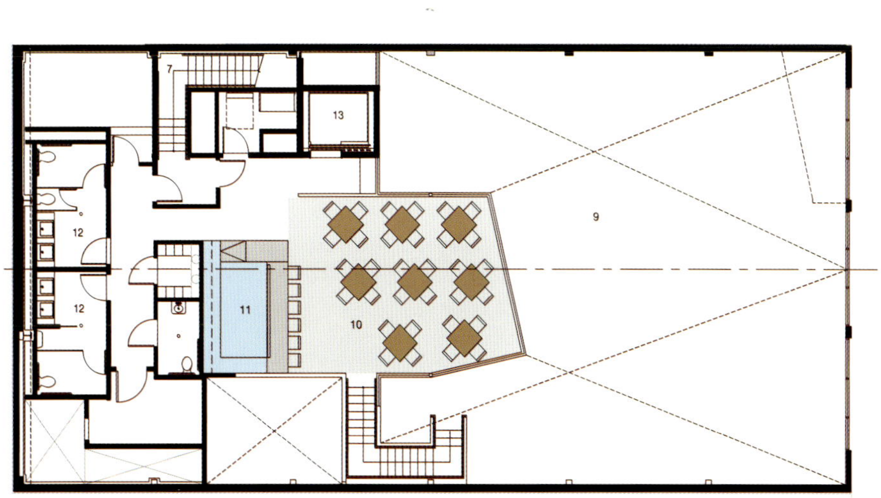

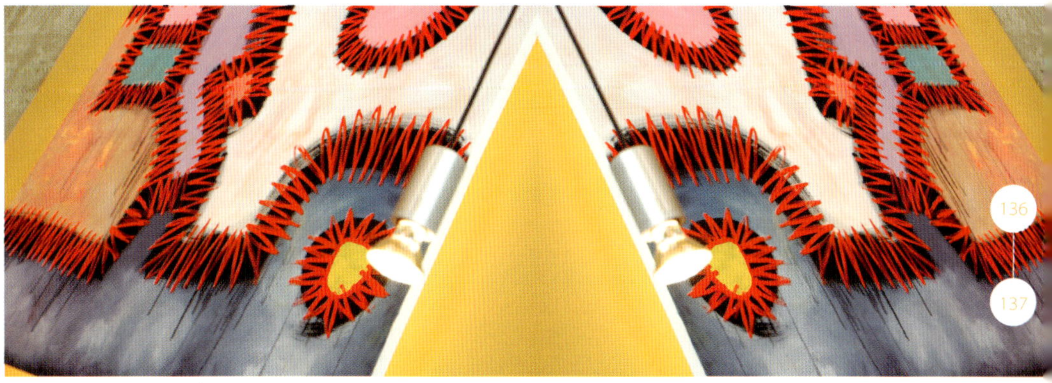

鲁比克餐吧

Project > 项目名称 > 鲁比克餐吧

Architect > 设计公司 > Arquitectos Interiores, S.C.P. 设计公司

Location > 项目地点 > 墨西哥

Built area > 项目面积 > 245 m^2

Photographer > 摄影师 > Eduardo Cervantes

项目介绍 PROJECT DESCRIPTION

Nested on the upper level of a shopping plaza, Rubik was conceived as a young but elegant eating place. Design was to develop an experience that built upon Merida's new wave of high quality gourmet restaurants.

鲁比克餐吧位于一家购物广场的楼上，是一家充满朝气而雅致的餐馆。它的设计继承了墨西哥梅里达地区高级美食餐厅的经验。

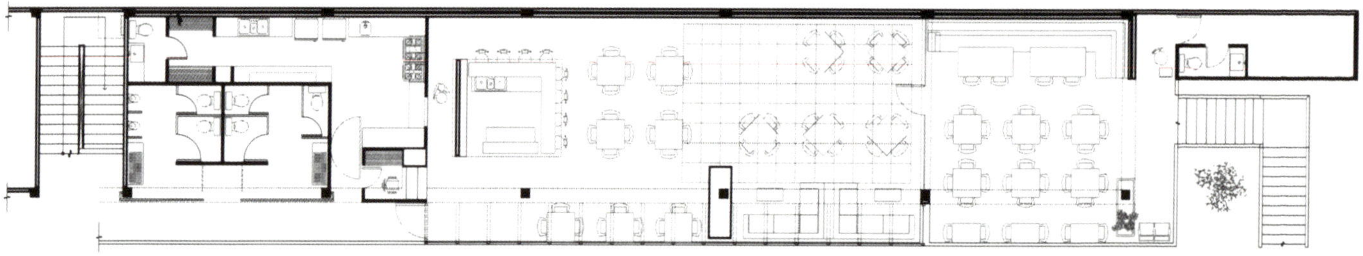

To attract clientele and make them feel comfortable, the space is outfitted with dark painted walls, carpet and porcelain floors and aluminum compound elements. The project maximized the environment by using the full width of the area and placing the bar as a focal point at the end corner concealing the service corridor and pulling the view to the back of the restaurant.

为了吸引顾客、制造舒适的氛围，整个空间采用深暗色墙漆、地毯、陶瓷地板及铝复合元素。该设计充分利用空间的宽度，将吧台放置于角落，作为空间的重点，隐藏了走廊，将整个视野拉到餐厅的尽头，使空间最大化。

rubik
restaurante bar
TODA LÍNEA COMIENZA EN UN PUNTO...
ESPÉRALO

EMPUJE

Diners can enjoy subtly different venues on every visit. A central dining area; more intimate clients may prefer a lounge seating located aside the main entrance. The most popular seats in the place are those on the front, a terraza-like area where you can drink or dine, either on a booth, table or standing alone with an unobstructed view to the city.

顾客每次到访都能感受不同的座位所带来的不同的微妙体验。有的顾客喜欢中央就餐区，而关系亲密的顾客可能更喜欢主入口旁的沙发座位。最受欢迎的座位是餐厅前面的阳台式区域里的座位：在这里，顾客可以品酒、进餐，可以选择单独的隔间或一张桌子，或者独自站在一旁，欣赏城市风景。

Muted shades of white and orange are used to color the dining area, while the bar is a bit brighter in tone. The metal feature counter on which the drinks are prepared serves as an oasis to its thirsty customers. Cocktail tables behind the bar seating stand on dark, single legs. Chairs, however, are on colored patterned upholstery.

用餐区域采用的是柔和的白色和橙色，相对于吧台来说，颜色更为淡雅一些。金属柜台为顾客提供了饮料，解其饥渴。吧台后面的鸡尾酒桌单脚站立，桌脚颜色为深暗色，而坐垫则采用彩色图案。

"城市中心"

Project > 项目名称 > "城市中心"

Architect > 设计公司 > Graft 设计公司

Location > 项目地点 > 美国拉斯维加斯

Built area > 项目面积 > 15810 m^2

项目介绍 PROJECT DESCRIPTION

City Center is a mixed use, 18 million square feet, development by MGM Mirage that features buildings by several world renowned architects.

"城市中心" 面积 15810 平方米，是一个由米高梅集团出资开发的多功能娱乐区，其中的建筑物由数位世界知名的建筑师倾心打造而成。

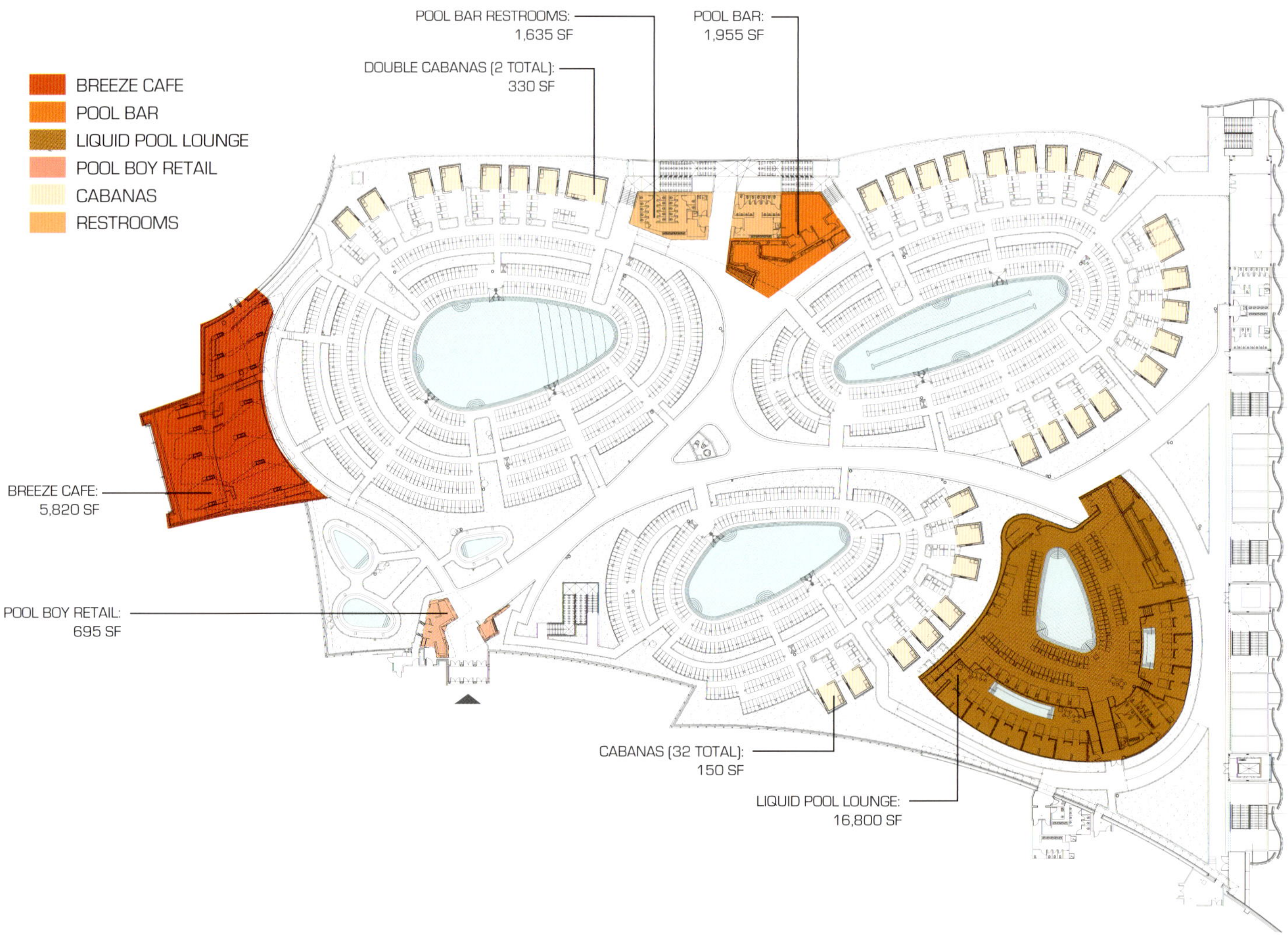

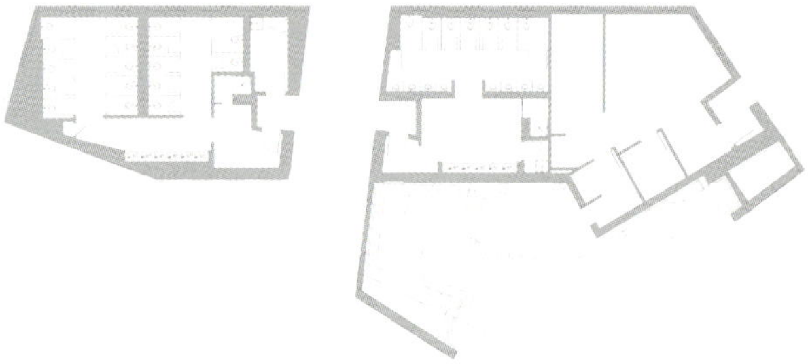

POOL BAR
1,955 SF

POOL BAR RESTROOMS
1,635 SF

The unique seventy-six acre urban resort is located in the heart of the famed Las Vegas Strip and is a collection of luxury hotels, condominiums, casinos, shopping facilities, and entertainment venues. To date, it is the largest privately financed development in the history of the United States and is also the largest LEED Certified project in the world.

这个无与伦比的 30 公顷的城市度假村位于著名的拉斯维加斯大道中心地带，聚集了豪华酒店、公寓、赌场、购物设施以及娱乐场所。迄今为止，此项目为美国历史上最大的个人投资项目，同时还是世界上最大的绿色建筑认证项目。

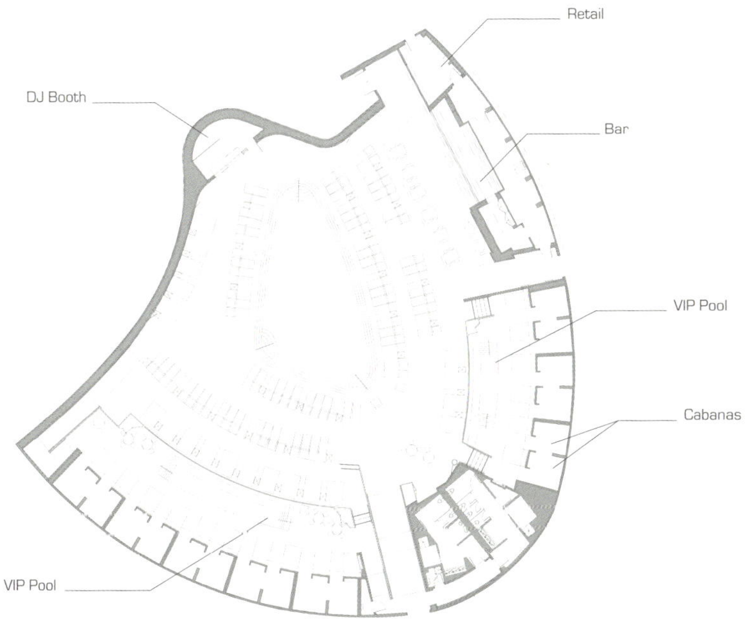

LIQUID POOL LOUNGE
16,800 SF

The pool bar is envisioned as a protected cliff dwelling in the center of the area. The large overhang provides cooling shade and protection from the sun. The monolithic white structure is opening itself to invite the guests, revealing a welcoming wood interior with an abstracted image of a canyon behind the bar.

湖畔酒家就像是湖畔区中心一处宁静的岩屋。巨大的屋顶可以遮阳，送上一片阴凉。其内部为舒适的木质构造，吧台后有一幅抽象的峡谷图像。这巨大的白色建筑正张开双臂，欢迎来自四方的宾朋。

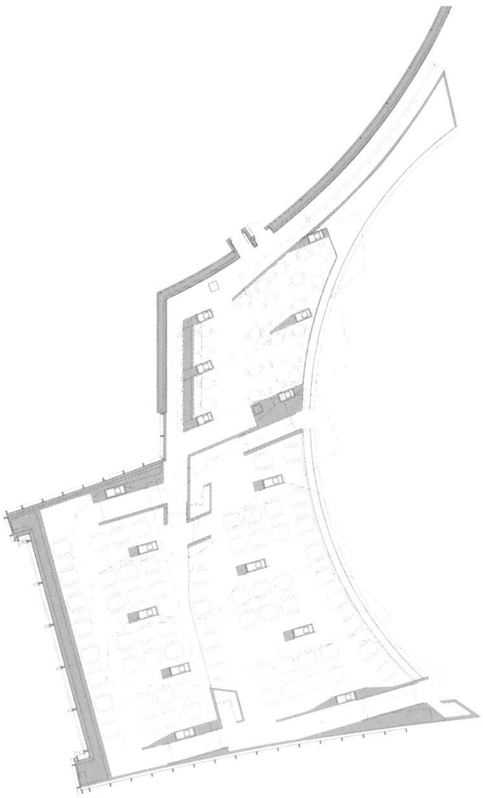

BREEZE CAFE
5,820 SF

The hungry patron can enter the Breeze Café and Pool Bar, located at the eastern end of the lagoon. The structure is inspired by the cooling experience of Mangrove forests. Large canopies provide shade, while simultaneously allowing free air flow through the openings between the roof structures, creating an enticing play of light and shadows in the space. The interplay of warm, comforting wood and sculptural white plaster surfaces enhances the experience of an opulent refuge from the sizzling sun.

清风咖啡馆和湖畔酒家位于环礁湖东侧，结构类似红树林，非常清新。巨大的天棚不仅可以提供一片阴凉，同时还可使空气在屋顶结构的空隙间自由穿行，创造出一种迷人的光与影的游戏。温暖舒适的木质表面与绘刻的白石膏立面搭配，更会带来别样体验，是躲避炎炎夏日的美妙去处。

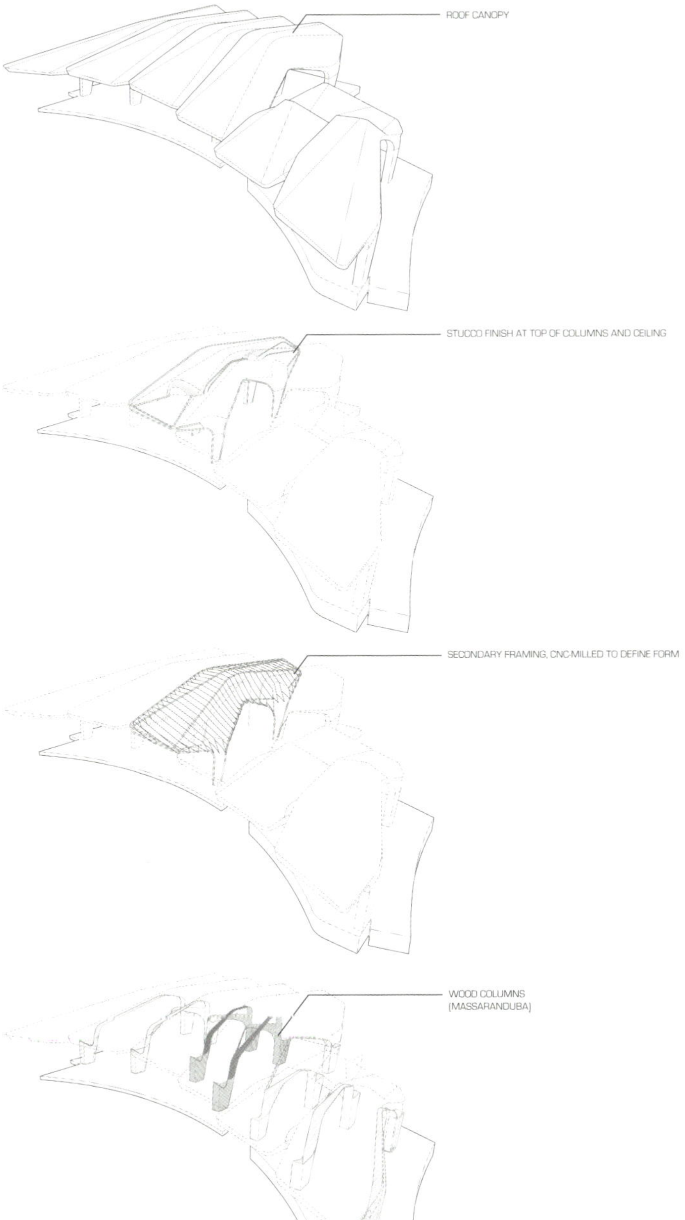

Breeze Cafe
Roof Structure

Cabana
Double Cabana

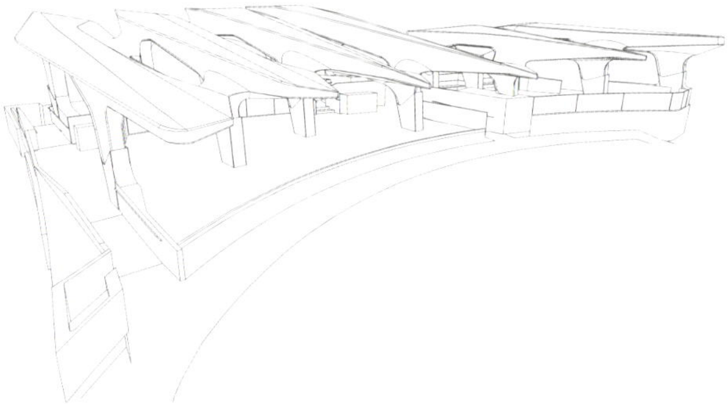

Breeze Cafe

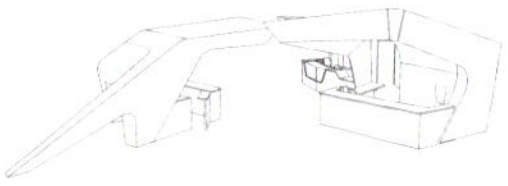

Pool Boy Retail

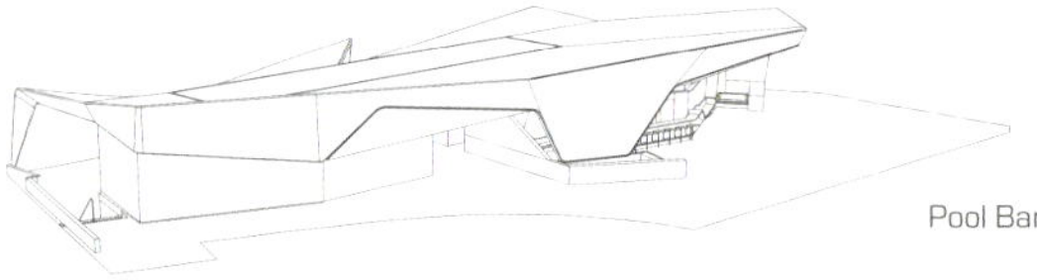

Pool Bar

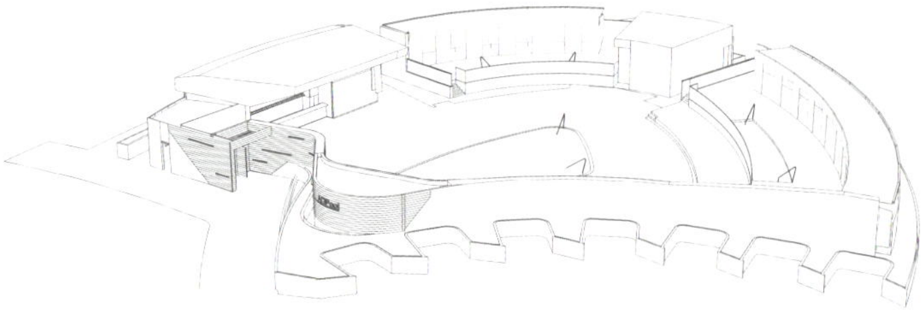

Liquid Pool Lounge

The third architectural component is a secluded cove at the western and of the pool area. It is an adult-only escape to another world of relaxation. A refuge was created that continues the material palette of the other zones, but uses a different architectural and spatial language, in order to set it apart from the adjacent areas. The Liquid Pool Lounge is operated by the Light Group. The back bar features original artwork by Los Angeles artist Camron Slocum.

第三个具建筑特色的去处为湖畔区西端一个幽静的港湾。此地仅为成人服务，可带您进入另一个休闲世界。这个幽闭之处延续了其他区域的材料风格，但为了与毗邻区域有所区别，又采用了不同的建筑语言及空间风格。后面的吧台陈列着拉斯维加斯艺术家卡姆伦 • 斯洛克姆制作的原始工艺品。

鲳参鱼餐厅

Project > 项目名称 > 鲳参鱼餐厅

Architect > 设计公司 > Studio Gaia 设计公司

Location > 项目地点 > 卡塔尔

Built area > 项目面积 > 200 m^2

studio GAIA architecture and interiors

项目介绍 PROJECT DESCRIPTION

This 200 m sq restaurant was conceived as an elegantly modern Mexican dining experience within the unique Qatari sands. Theatrical tapered columns with Mexican motifs soar 4.5 m high to create a dramatic yet subtle ambiance. Taking our cue from the local environment, we selected muted, sandy tones of upholstery, plastered walls and polished sandstone floors to create a warm and comfortable setting.

这个200平方米的餐馆被设计成卡塔尔沙漠中一座高雅而现代的墨西哥餐厅。几根4.5米高带有墨西哥图案的锥形立柱，创造了一个具有戏剧性且精致无比的环境。基于当地情况，我们选择了柔和的沙土色调软装、泥灰墙和抛光的砂岩地面，创造出一个温暖舒适的氛围。

The name 'Pampano' refers to a delicious fish, and the fish motif is featured everywhere, with artful black and white images playfully strewn across the exterior windows and mirrors by Mexican photographer. A dramatic 3 m screen above the bar is composed of schools of tiny bronze fish, commanding the bar area. The bar itself was designed with human interaction in mind, as the sunken bar floor allows seated patrons to interact with the bartender at the same eye level for a more intimate experience. Behind this, a waterfall trickles down a ruby red and black glass mosaic wall to lend a punch of color.

"鲳参鱼"是一种美味的鱼，而这种鱼的图案在餐馆中随处可见，设计师将其刻画成黑白的艺术形象，让其穿梭在墨西哥摄影师设计的外窗和镜子之间，妙趣横生。吧台上方还有一个戏剧性的三米屏幕，由无数的小铜鱼构成，装饰了整个吧台。吧台设计利于人与人之间的交流，地板平面相对较低是为了让坐着的顾客能更好地与酒保进行面对面的接触。吧台后面，是一个红宝石与黑玻璃马赛克墙，瀑布顺其流下，给顾客一种丰富多彩的视觉体验。

A swirling, 16 m long feature wall evokes the shifting sands of the Sahara desert and provides an eye catching backdrop for the abstract sculptures of Mexican artist Paloma Torres. Meanwhile, circular banquettes and sheer drapery partitions allow diners to have an intimate eating experience.

一个 16 米长的漩涡式的景观墙，展示了卡塔尔沙漠沙层变化的美妙。墨西哥艺术家帕洛马托雷斯的抽象雕塑作为背景墙，着实引人注目。此外，圆形的长凳和透明的坐垫能让顾客享受亲切的就餐体验。

We designed the orchestra-inspired maitre'd stand as a whimsical tribute to co-owner Placido Domingo, the legendary opera maestro. Climbing up a flight of stairs leads you to an intimate private dining room, with wine storage and privileged views overlooking the bar.

管弦乐队式的领班台的设计是为了纪念传奇歌剧大师兼餐厅的合资老板普拉西多 • 多明戈。爬上一段陡峭的楼梯之后就进入了一个私密的就餐房间，里面储藏着大量的葡萄佳酿。

RICETTA LIETO 餐厅

Project > 项目名称 > RICETTA LIETO 餐厅

Architect > 设计公司 > ICHIRO NISHIWAKI 设计公司

Designer > 设计师 > Ichiro Nishiwaki

Photographer > 摄影师 > Nacasa&Partners Inc

项目介绍 PROJECT DESCRIPTION

It's a casual restaurant for French and Italian cuisine. The restaurant is in the basement of the commercial complex environment and its space has a tall ceiling and a narrow width. Seats are arranged in a single direction and the space is added an impressive impact with its mirror effect and the originally colored construction of surface materials of multiple stripes along its orientation.

RICETTA LIETO 餐厅是一家经营法国菜和意大利菜的餐厅。它位于一家综合性商厦的底层，天花板很高，空间较窄。餐厅的座位简单按列排开，空间内的镜像效果令人印象深刻。

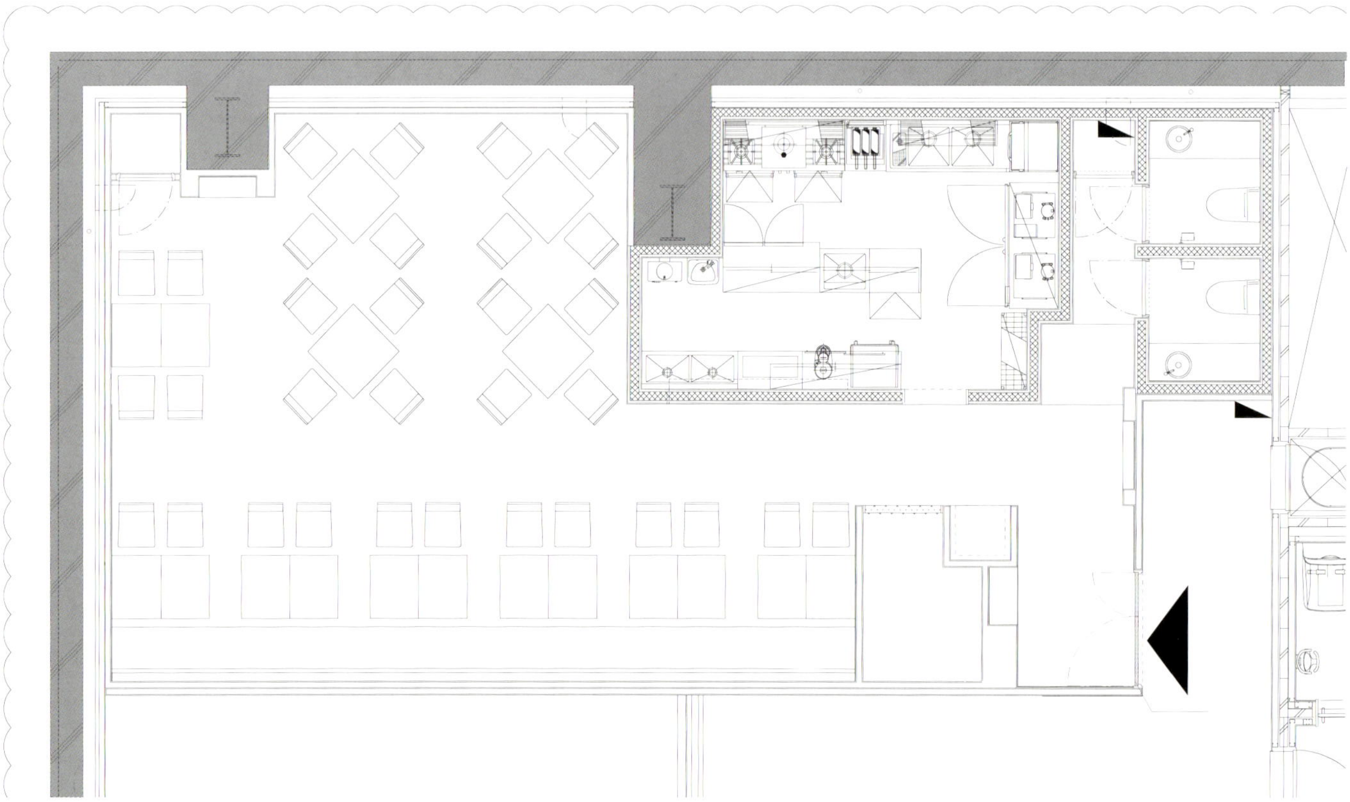

Entrance to the restaurant is in the dark colors and it brings a mysterious feeling to guests. Through the small window on the wall, the scene of inside can be seen. There are four photographs on both sides of the window, symmetrical arrangement of them are in cooperation with the placing of the seat. Wall building material has the inherent natural colors according to the location of the space is divided into several zones.

入口处的色调偏暗，增添了一种神秘色彩。透过墙壁上的小窗子可以看到餐厅内的景象，窗子两侧分别有四幅照片，对称的排列与空间内座椅的摆放相呼应。墙面材质所具有的建筑固有色调根据固有的位置将空间分为多个区域。

RICETTA LIETO
Italian & French

In addition, the space is furnished with a live atmosphere by the indirect light from the upper part of high back seats and the down light from crystal balls under the ceiling. The ceiling above the seats with a colored stripe pattern, crystal pendant hanging on the ceiling reflect the light, and in cooperation with the colorful patterns on the ceiling to make the symmetrical space more varied. Black and white photos on the walls are in contrast with the color of the space highlighting the restaurant features.

高靠背椅子上方的间接照明和天花板下面的水晶球发射的灯光使餐厅的氛围活跃生动。座位上方的天花板采用了彩色条纹图案，天花板上垂下的水晶吊饰反射了灯光，而且天花板上多彩的图案使对称的空间更具多变性。墙壁上黑白色的照片与空间的色调形成对比，突显了餐厅的特点。

Praq 餐厅

Project > 项目名称 > Praq 餐厅

Architect > 设计公司 > Tjep. 设计公司

Location > 项目地点 > 荷兰

项目介绍 PROJECT DESCRIPTION

Two wonderful entrepreneurs came to us with their wish to start a new unique restaurant concept that would welcome parents and their children without looking like a playground. Indeed parents don't necessarily want to spend the evening with Donald Duck when they go out (nor do the children by the way!). The first Praq restaurant opened its doors shortly after and became quite successful and well-known.

两个热忱、想法独特的企业家带着他们新奇的构思找到了设计师，他们带来的是一个独一无二的、全新的餐厅理念——一个家长和孩子都向往的餐厅，却又不像游乐场所。确实，父母们都不希望自己晚上出去消遣的时候跟唐老鸭呆在一起（顺便提一下，孩子们也抱着同样的想法！）。因此，第一家 Praq 餐厅就在这之后很快诞生了，并获得成功，也逐渐家喻户晓。

- •1 bar
- •2 kitchen
- •3 kids + parents space
- •4 growen-up space
- •5 balancing dish table
- •6 tunnel benches
- •7 window table
- •8 car table
- •9 bus table
- •10 movie house
- •11 drawing table
- •12 kitchen table
- •13 game table
- •14 toilets

After 5 successful years our client realized that the formula had the potential to become a restaurant chain, thus opening two new venues in 2008. Again we tried to create a restaurant that is inspiring and fun for children, while being comfortable and enjoyable for their parents. This is in an integrated rather than separated manner, which is more usually the case.

在五年的成功经营之后，客户意识到运用这套经营理念完全有潜力发展成连锁企业，随即08年两家餐厅旗舰店也相继诞生。再次强调一下，这是一家对孩子既具有启发，又有乐趣，同时对父母来说舒适、轻松的餐厅。这是一个复合型的创意，而不是普通的单一风格。

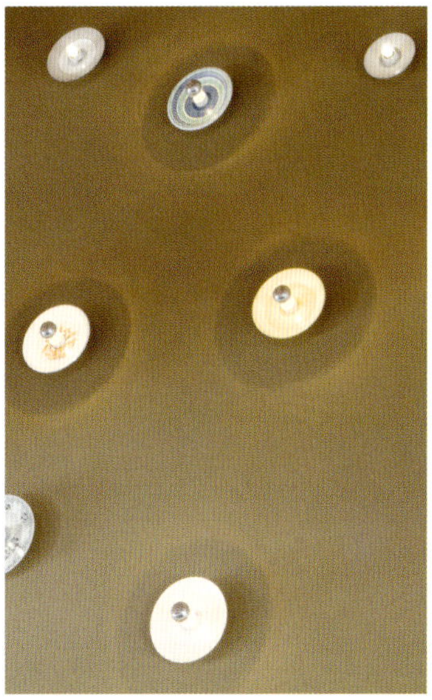

One space is reserved for children and their parents while the other is reserved for adults. The adult space clearly fits the same world, only here the accent is more geared towards cosiness and style. By the way Praq means mashed food for children in Dutch, however, the plates served at Praq are very refined and tasty.

餐厅一部分空间专为父母和孩子设计，一部分专为成人设计。成人的空间与外界并无很大的差异；只是，这里的风格更倾向于舒适安逸，具有设计感。顺便提一下，Praq 在荷兰语中指的是适用于儿童的捣碎的食物，而在 Praq 所提供的菜肴都十分精致美味。

The space is characterized by a monumental farm style roof composed of huge massive wooden beams. Within this space we tried to create a playful world by placing furniture and elements that form crossovers between objects and furniture or furniture that refers to the space itself. For example a table becomes a window, a bus or a kitchen. The six meter high construction in the center looks like an abstract cloud evoking something of a colorful game while contrasting nicely with the handcrafted architecture.

餐厅的建筑特色是一个由庞大的木梁搭成的农场屋顶。在餐厅内部，我们试着通过铺设家具和一些在物体、家具或空间本身之间形成的交叉元素来创造一个生动有趣的世界。比如，将一张桌子变成一个窗户、一辆巴士或者一个厨房。中央位置六米高的结构看起来像一团抽象的云雾，五颜六色，同时又与手工搭成的精美建筑物形成和谐对比。

Murasaki 餐厅

Project > 项目名称 > Murasaki 餐厅

Architect > 设计公司 > Infix 设计公司 （Yoshihiko Mamiya）

Location > 项目地点 > 中国上海

Built area > 项目面积 > 385 m^2

Photographer > 摄影师 > Nacasa&Partners

项目介绍 PROJECT DESCRIPTION

"Murasaki" is the first-class Japanese restaurant which opened temporary in a Japanese industrial pavilion in the Shanghai international exposition. First of all, gardens are inevitable for the first-class Japanese restaurants. In this project, the garden with the spirit of Japanese culture such as subtle taste and elegant simplicity, fugacity and incomplete aesthetics embodied, is the highest priority to create.

Murasaki 是上海世博会日本工业馆内临时建立的一家高级的日本餐厅。餐厅内的花园是必不可少的，它是日本文化的反映。精细的品味和简单、多变而又不失高雅的设计呈现出日本文化残缺美的审美观，这是设计首先要表达的。

But it is not a traditional formal garden made of soils. In addition, they planed to provide visitors with calm time to meditate when watching the garden isolated from the noise in the exposition like a mountain lodge in a city. Moreover, in recent years, what exhibited in international expositions almost appeals to visual and auditory perception, not concrete matters and lack of materiality. They aimed to restore the materiality in this restaurant in opposition to the recent trend.

这里的花园不是传统意义上的土质花园。设计师想为参观者创造一个静谧的环境，在这里参观者可以不受打扰地进行思考，欣赏花园美景，似乎身处城市中的山脉，远离了外界的喧嚣。近年来在博览会上展出的多为视觉和听觉上的元素而非具体的物质，缺乏物质性。设计师想更多地在这个餐厅中还原物质性，而不跟随近期的趋势。

料亭
紫
MURASAKI

In five private rooms, the yukimi-shojis are door, window or room dividers consisting of translucent paper over a frame of wood which holds together a lattice of wood or bamboo. The lower part is glass unifying the interior with the view of the garden, which evokes viewers' imagination.

五个包间的门、窗和房间的隔断都用了yukimi 障子，木框上糊了半透明的纸，木框多为木板框架或竹架，下面的部分是玻璃，将室内与花园连接起来，激发参观者的想象力。

In the approaches to the private rooms, natural phenomena were created on the walls by weaving thousands of wires which was likened to rain represented on Ukiyoe. Although this space disappeared in the end of the Shanghai international exposition, it kept high quality transcending an immediate pavilion. It was the Japanese aesthetic sense in which people admire the changes.

通往包间的墙面上用许多的线编织成网，自然地形成如浮世绘中描绘的从天而落的雨滴的景象。尽管上海世博会后餐厅被拆除了，但是它却是一个超越了临时性展馆的优秀作品。参观者可以从中体会到日本文化的审美感，感叹设计的多变和精巧。

Then, they made the garden in cooperation with art curators by composing of art pieces of two artists. The two artists are Fumiko Nakatani and Makoto Azuma who expresses fugacity and incomplete aesthetics in their works. As a result, the garden had their works add dignity to the whole restaurant while Japanese aesthetic sense was embodied in the interior.

设计师与艺术馆馆长借出两名艺术家的作品来装饰花园。两名设计师分别是 Fumiko Nakatani 和 Makoto Azuma，他们的作品也表达出多变、残缺美的审美观。在装饰了艺术品后，花园为整个餐厅增添了高贵气息，内部设计也凸显了日本文化审美感。

玛莉索餐厅

Project > 项目名称 > 玛莉索餐厅

Architect > 设计公司 > VA

Location > 项目地点 > 西班牙

Built area > 项目面积 > 596 m^2

Photographer > 摄影师 > Jose Manuel Cutillas

ms

LOUNGE MS - CADREITA

项目介绍 PROJECT DESCRIPTION

Marisol restaurant proposes the construction of a new space as an annex to its existing facilities to accommodate additional uses to events, the need for a covered space which is eminently open and differentiating environments without losing their specificity of terrace, bar, lounge, TV and dance...

玛莉索餐厅创造了一个新空间作为原空间的附属建筑，可以承担一些 party 或者活动，业主要求这个有顶棚的空间，必须开放，且隔出露台、酒吧、休息室、舞蹈间等功能区。

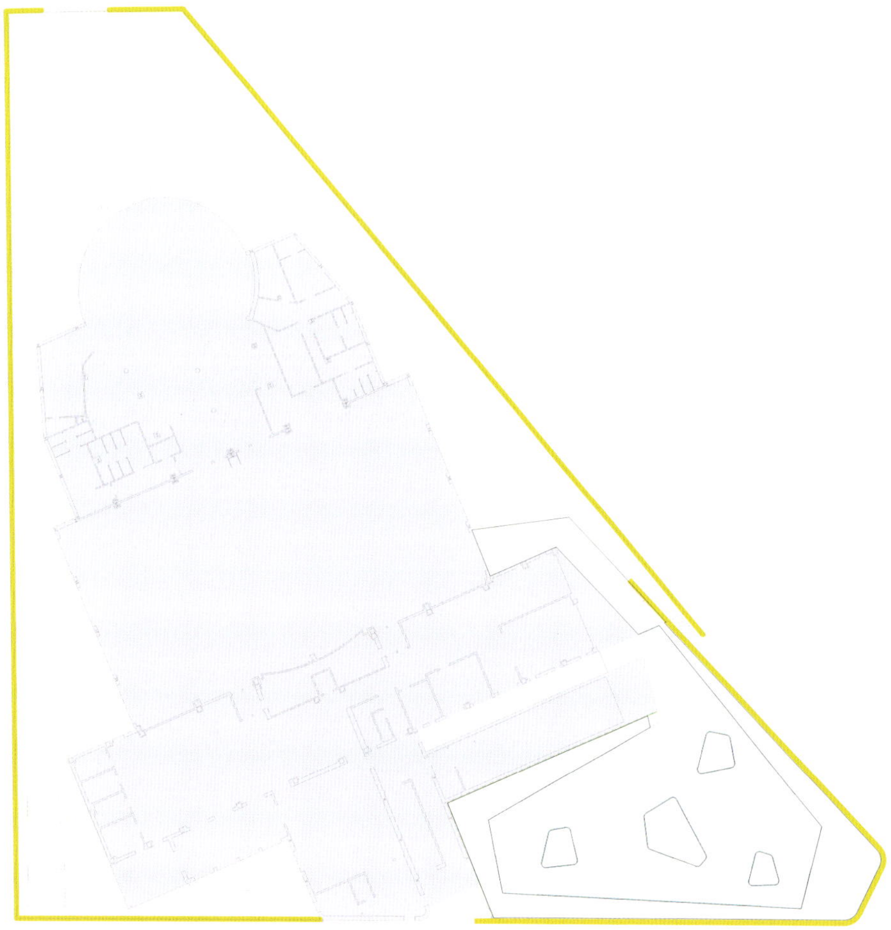

SITUACIÓN e:1/400

This species builds a base of recycled plastic tubes of different colors similar to reeds, organizing a braided flexible and deformable organic capability to adapt to any situation and geometry. This new plant species proposes a closure permeable, breathable, which acts as both solar controls, thermal and visual filter.

整个建筑的外立面是由看似芦苇一样的各色环保塑料管组成，编织出一个柔韧度极强的有机体，可以根据情况变形成任何几何形状。这种新的植物可以制造具有透气和可呼吸的围合空间，同时可以控制外部的太阳能、热能，变化视觉效果。

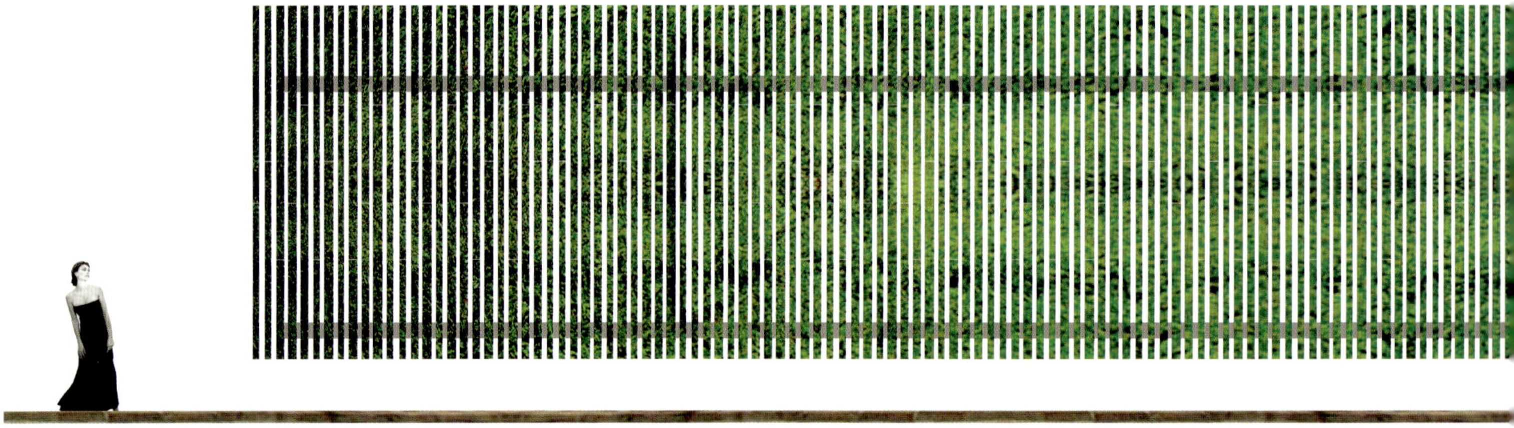

DETALLE FACHADA

01 LOSA e=20cm CANTOS Y CARA INFERIOR VISTA ACABADO DELTA DRAIN GRAN FORMATO
02 JUNTA DE NEOPRENO. ESPESOR Y DENSIDAD S/ESTRUCTURA
03 PLACA ANCLAJE ENRASADA CON LOSA DIMENSIONES Y ANCLAJE S/ESTRCUTURA
04 TAPAJUNTAS ALUMIO ANODIZADO 20+100+20 e=2mm
05 VIDRIO LAMINADO CON BUTIRAL TRANSPARENTE 8+8
06 CARPINTERÍA CORREDERA: ALUMINIO ANODIZADO. DE LA SERIE 3001 DE VITROCSA.
07 REJILLA LINEAL
08 EMBELLECEDOR PILARES/BAJANTES ACERO CORTEN e=1mm Ø14cm h=20mm
09 SOLERA HORMIGÓN PULIDA e=10cm
10 LOSA CIMENTACION e=25cm CON CANTOS VISTOS
11 REGULADOR DE CAUDAL Ø122
12 TABLERO PERFORADO S/CLIMATIZACION 45x130x2 COMPLETAMENTE SELLADO
13 HORMIGÓN DE LIMPIEZA e=5cm
14 TODOUNO COMPACTADO ESPESOR S/ESTRUCTURA
15 CONDUCTO IMPULSION 1500x400 MURO H.A. VISTO CARA INTERIOR 15cm
16 GRAVA
17 SUSTRATO RESISTENTE COMPACTADO

01 REMATE DE CHAPA DE ZINC CLIPADO A FLEJE 100+40+140+260+25+20 = 585mm
02 CAZOLETA CON PARAHOJAS
03 REFUERZO IMPERMEABILIZACION SOBRE IMPRIMACIÓN BITUMINOSA
04 JUNTA ELASTICA 1cm
05 BAJANTE ACERO GALVANIZADO Ø63.2
06 SOLERA HORMIGÓN PULIDA e=10cm
07 LAMINA IMPERMEABILIZANTE
08 COLECTOR PVC S/SANEAMIENTO
09 PAVIMENTO URBANIZACION BALDOSA HIDRÁULICA SOBRE SOLERA H.A. e=15cm
10 LAMINA IMPERMEABILIZANTE
11 DREN LINEAL Ø160
12 JUNTA PVC
13 ZAPATA CORRIDA
14 HORMIGÓN DE LIMPIEZA e=10cm
15 SUSTRATO RESISTENTE COMPACTADO

The new space is conceived as a continuation of the existing fence, wrapping it all, but hidden: expressing their new identity, not as constructed element, but as a re-forested. A new plant species grown in the surrounding area … the new scalar similarity between elements of "plant" makes the proposal to a new understanding of the link with the existing connection.

新的内部空间被做成是外立面围墙的延续，围墙将整个建筑包裹，并露出一部分，用植被一样的装饰而非建筑元素来充当空间标志。内外各种"植物"元素之间的照应，将整个空间统一起来。

The surfaces of glass and concrete walls that make up the project have a poor acoustic performance. This is why we have worked the surface of the slab covered by a system of "egg-crate". This texturing makes the wave break and reduce the reverberation of the room. It overlaps on each round wood roofs in areas with higher noise level.

由混凝土、玻璃和混凝土墙构成项目的表面隔音效果通常都很差。这就是为什么我们要在一个墙面上覆盖“蛋箱”一样的条板结构，阻止声音泄露。这种纹理也能阻挡声波，减少房间的混响，并且能在圆形木屋顶区域将声音重叠，达到更好的音效。

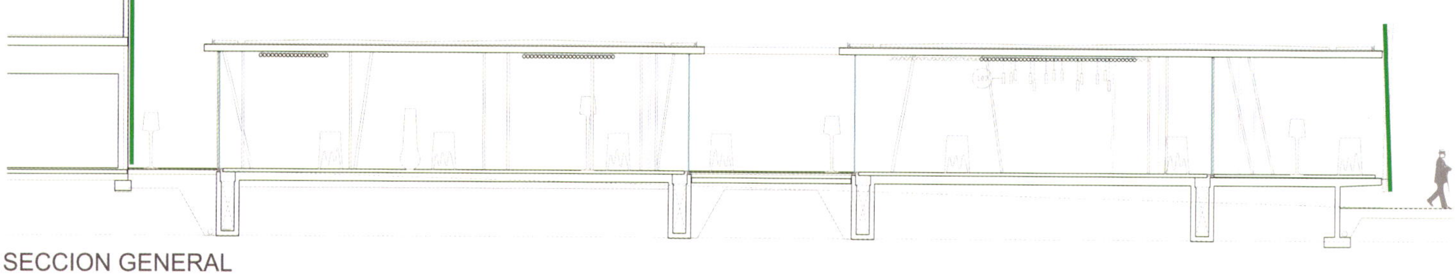

SECCION GENERAL

The organizational scheme is due to similar patterns of micro-structures, more in line with geometric patterns of liquid and/or aerosols that Cartesian structures in establishing a working geometry using "soft" and a unified treatment space "airy".

整个建筑结构取材于微结构，同流体的几何图案或者同笛卡尔结构中使用的"柔软"的几何图形和"空灵"空间的手法一致。

The thermal shutdown is achieved by a laminated curved plane glass. The lower bearing system enables the sliding glass supported directly by these bearings. Carpentry, therefore, is reduced to a profile that protects the edge of the glass.

整个空间靠一块弯曲的夹层玻璃来隔绝外界的热气。低矮的系统直接托起滑动玻璃。这样，木器就保护了玻璃的边缘。

Inamo 餐厅

Project > 项目名称 > Inamo 餐厅

Architect > 设计公司 > Blacksheep 设计公司

Location > 项目地点 > 英国伦敦

Built area > 项目面积 > 310 m^2

项目介绍 PROJECT DESCRIPTION

Inamo, a whole new concept in west end dining, opens this month in the heart of London's Soho, with menus projected onto tabletops, allowing diners to order food and beverages interactively, to change the ambiance of their individual table, to play games and even to order up local information and services, such as booking a cab! The 310 sq m (gross), 62-cover restaurant will offer customers high-quality Asian fusion cuisine.

Inamo 餐厅是伦敦餐饮界的一家创新性的餐厅，位于伦敦 SOHO 区。它最有特色的地方在于将菜单投影在桌面上，顾客可以进行互动式的点餐，还可以根据需求改变自己桌子周围的环境，或是在桌面上玩桌游打发时间，甚至可以通过这里了解当地的信息、订购服务，比如可以在饭后叫一辆计程车。面积约 310 平方米、拥有 62 张"点餐餐桌"的 Inamo 餐厅，给顾客带来高科技体验的同时也带来了别具风情的亚洲餐饮艺术。

Customers approach the restaurant via a brand new jet black timber facade, in keeping with the existing architecture. To protect interior light levels, the glazed restaurant windows have a graphic vinyl applied featuring solid and translucent star patterns, creating a strong graphic statement as well as protecting interior light levels. The vinyl graphic nonetheless allows tantalising glimpses of the interior and of the Blacksheep-designed 'cocoons' which house the projectors, computers and frames and which sit above each table.

顾客绕过正面黑色木板做成的墙面可进入餐厅。为了保证室内充足的照明，餐厅的玻璃应用了固态乙烯基和半透明的星形组合，给人一种强烈图形感的同时也保证了室内的照明。这些图案不仅增强了室内的观赏性，也装饰桌子上方外形像蚕茧的投影仪。

inamo

inamo
drinks
inamo

The central entrance to Inamo is via glazed doors with a high-impact, red, black and white graphic applied to the back wall of the small entrance lobby area. Once inside the restaurant, the stair to the downstairs bar is immediately to the left, finished in matt black, with timber-effect vinyl flooring here and throughout. Graphic wallpaper above the stair mirrors the window graphic. The restaurant has been treated partly in monochrome to allow the tabletop illuminations to stand out and partly with further graphics (inspired by the colour choices of kaleidoscopes and origami for an Asian feel) to mirror the strong graphic sense created by the tabletops.

餐厅中间的过道安装有玻璃门，上面是具有强烈视觉冲突的红色图案。通往休息区的入口处墙面上有黑白色图案。通过餐厅里的楼梯可下到底层的吧台，楼梯是亚黑色，木材质感的乙烯材料贯穿始终。楼梯上的图案与窗子上的图案相呼应。餐厅的部分空间采用单色调以突显桌面的灯光效果，另一部分应用了更复杂的图案，这来源于亚洲特色的万花筒色彩元素和折纸艺术，带来了强烈的图形感。

The restaurant design had to have an overall strong sense of identity as a space, neither overwhelming nor being overwhelmed by the technology at its heart. "We created a concept based around the keywords: warmth, vibrancy, charm and theatre", explained Blacksheep Director Tim Mutton."It was important to balance out the restaurant's technological aspects by creating a sensuous, social space with a strong personality."

餐厅的设计具有强烈的空间感，既不能被技术主导也不能将技术特点掩盖。"我们设计的理念基于以下几点：热烈、活力、魅力与戏剧化。" Blacksheep 公司的负责人 Tim Mutton 说，"使用个性强烈的风格，努力创造一个社会化的空间来平衡餐厅的科技化元素是非常重要的。"

O2 餐厅

Project > 项目名称 > O2 餐厅

Architect > 设计公司 > Garduño Arquitectos （Juan Garduño）

Location > 项目地点 > 墨西哥

Built area > 项目面积 > 930 m^2

Photographer > 摄影师 > Laura Cohen、Sófocles Hernández

项目介绍 PROJECT DESCRIPTION

O2 is the conversion underground plaza for a parking lot into a garden restaurant with a new approach and ingenious sustainable ideas. When designers were confronted with the challenge of designing a restaurant in an overcrowded and concrete-saturated area, they set out to create something that would also make a contribution to area, with vegetation, warmth and a commitment to improving the immediate surroundings.

O2 餐厅是由一个位于下沉式广场的停车厂改造成的花园式餐厅，设计手法新颖而独具匠心。设计师面对的是一个繁华拥挤、被钢筋混凝土充斥的区域，设计方案同时也会给这里带来一些改变，植被的增加以及给周围环境带来的温暖、亲切的氛围。

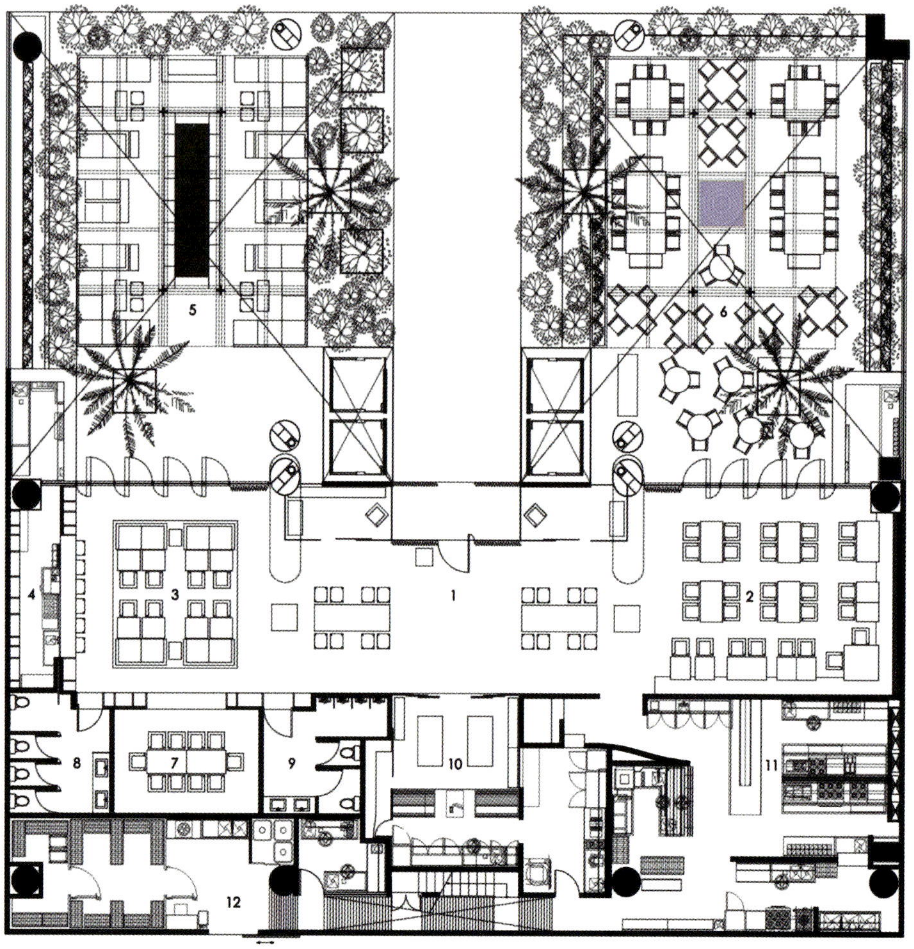

1 Reception
2 Dining room
3 Interior lounge
4 Bar
5 Lounge / Terrace
6 Dining / Terrace
7 Private room
8 Bathroom - Ladies
9 Bathroom - Gentlemen
10 Deli store
11 Kitchen
12 Service Access

Restaurant O2, A garden in the parking lot!
Mexico City
Garduño Arquitectos

The proposal revolved around sustainability and a connection with contemporary art. It includes a restaurant with terraces, fountains, chimneys and sitting areas, a gourmet store, jewelry shop, bakery and flower shop all in 930 square meters. Both terraces are immersed in a microclimate of diverse plant life: a "designer garden". They were built around a central chimney, a fountain and watered through micro-nebulization.

设计师更多地考虑到可持续性，也在设计中融入了现代艺术元素。在930平方米的范围内，餐厅包含了露台、喷泉、雅座、烟囱、美食商店、珠宝店、面包店和花店。楼台上种满了各式各样的植物，被称为"设计师的花园"。植物围绕中心烟囱和喷泉，通过微型气化设备灌溉。

Some sections use LED lighting and the furniture is made of certified wood. The floors and bookshelves are made of reused pine planks, and all toilets have water saving fixtures.

一些区域采用 LED 照明灯，家具均为木质家具。地板和书架采用再生松木板，卫生间均安装有节水设备。

0% pet

you & me

LOVE

A unique heating system was invented for this place, where the heat generated by the kitchen is absorbed by copper coils containing a fluid with a very high boiling point. After absorbing the heat as it passes by stoves and extraction hoods, it leaves the circuit at a temperature of 40 degrees Celsius. The heat is then channeled through the terrace floor and to the bases of masonry couches in the terrace bar, heating the area for the comfort of clients and freeing the kitchen from using an additional mechanical cooling system.

餐厅有完善的供暖系统，厨房散发出的热量被铜卷管吸收，里面装有高沸点的液体。吸收后的热量被传递到炉子和排气罩处，这样的循环使温度上升到 40 度。热量随后通过露台的地板传送到吧台石椅下面的区域，使这里的温度适宜，厨房也因此不需要再安装降温设备。

An additional proposal was to eliminate bottled water so as not to contribute to the ecological costs of PET. For this purpose, we built a water purifying plant, which is visible on the terrace, and we designed a water bottle made of a recycled wine bottle. The restaurant only sells its own water and reuses its bottles. All proceeds from the sale of the water are destined to charitable causes. Every disposable product used is biodegradable, including the paper used for printing menus, bags and straws.

为了减少对瓶装水的需求以降低 PET 的生态成本，设计师在露台设计了一套净水设备，同时还设计了用可回收酒瓶做成的水瓶。餐厅只销售自己的水，回收自己的瓶子。所有收入均捐给慈善机构。所有的一次性产品均可降解处理，包括用于菜单的纸张、纸袋和吸管。

Vesu 餐厅

Project > 项目名称 > Vesu 餐厅

Architect > 设计公司 > Arcsine Architecture 设计公司、Bellusci Design 设计公司

Location > 项目地点 > 美国加州

Photographer > 摄影师 > Sharon Risedorph

项目介绍 PROJECT DESCRIPTION

Arcsine Architecture and Bellusci Design transformed a piano store with little curb appeal into a vibrant and engaging restaurant, fitting the bustling downtown Walnut Creek. The clients sought an inviting, modern environment to complement the globally inspired, locally sourced menu by Executive Chef Robert Sapirman. Situated on a competitive street corner, the design needed to grab attention and engage passersby.

设计师将一家路边的钢琴店变成一家有活力而迷人的餐厅，与闹市区的浮躁相统一，业主希望为客户提供一个诱人而现代化的就餐环境，以配合餐厅的特点 —— 将全球的时尚元素融入到本地化的食谱之中。该餐厅位于街角，所做的设计必须要能吸引过往行人的注意。

vesu
1388

vesu
vesu

vesu

The exterior wood panel rainscreen system is Prodema ProdEX. The rounded nature at the corner responded to context in downtown Walnut Creek – many of the buildings have radiused corners – but not in such a contemporary manner. Further wrapping wood, glass and anodized aluminum felt like a nice way to push these materials slightly.

外立面的木板防水系统采用 Prodema ProdEX。因为当地许多建筑都是弧形拐角的，所以此建筑的外观同当地环境相吻合，又显得十分现代。此外，表面的木材、玻璃和氧化铝材质更加对比突出这种感觉。

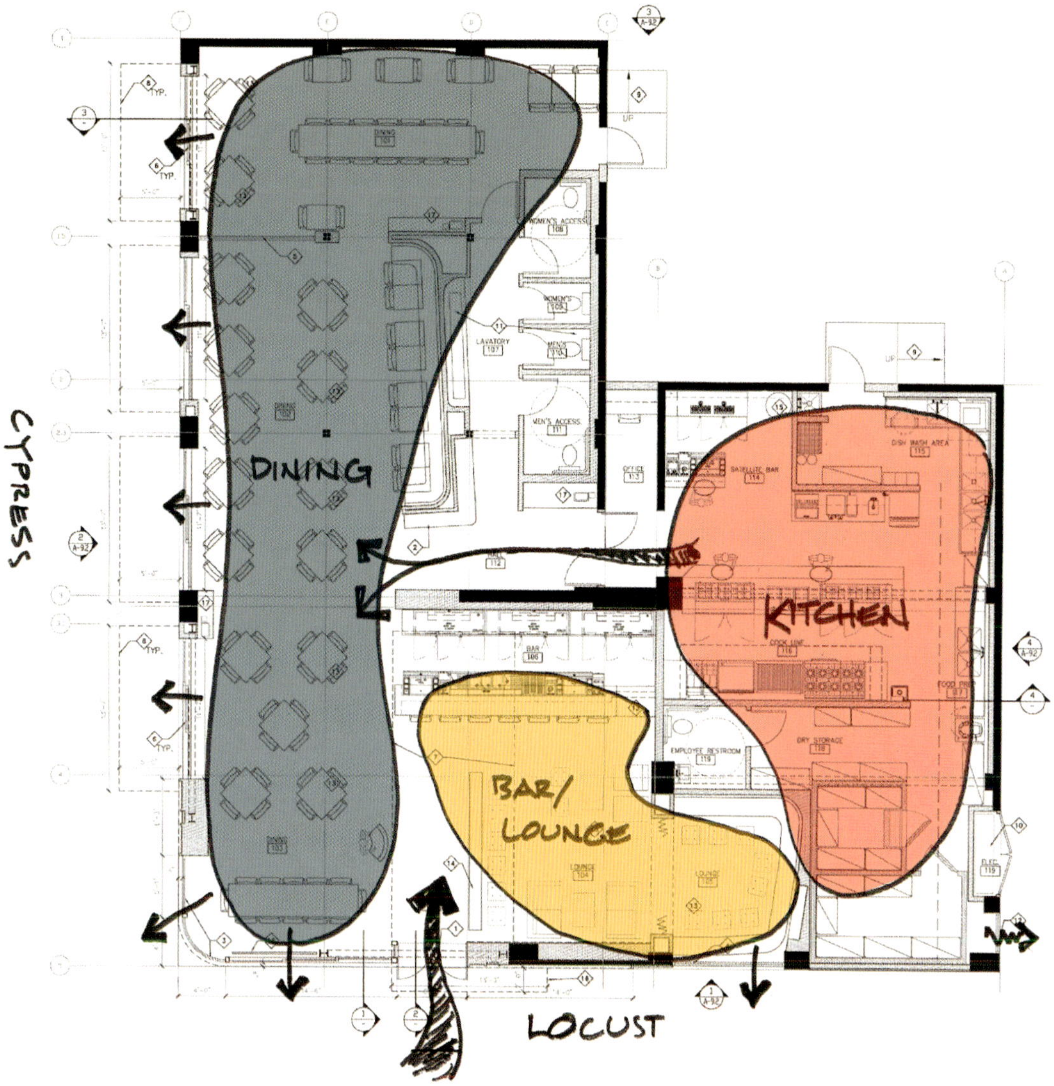
DINING
KITCHEN
BAR/
LOUNGE
CYPRESS
LOCUST

vesu

Arcsine wrapped the building with a contemporary, wood-panel façade, complimented by a curving wall of frameless glass that entices pedestrians to look in and feel as if they are already inside. An Arcsine-designed canopy of FSC Certified Eucalyptus leads patrons from the entrance to the lounge and extends all the way to the back bar.

建筑的立面采用现代风格的木板装饰，令人称赞的是无框玻璃曲面墙的设计，方便行人看到室内情形，仿佛置身其中。带有浓厚设计感的吧台天棚使用通过 FSC 认证的桉木材料，将顾客从入口处引入到后面吧台的休闲区。

The Eucalyptus awning became a great way to capture attention to the entry and signage and also offer some comfortable compression as one walks into the space. By letting this awning soffit complete an "S-curve" to a higher point, it opens up the space at the lounge area. Allowing it to curve down at the back bar permits the exquisite grain and warmth of the wood to become the icon of the restaurant and lounge experience.

入口处的桉木天棚是一种吸引眼球的好方法，成为该餐厅的标志，也为进入此空间的人们提供舒适的感觉。此拱顶最高处形"S形曲线"，沿曲线往下延伸到后面的酒吧，扩大了休闲区。美味的食物和温暖的木质结构成为该餐厅的标志和休闲乐趣所在。

Bellusci Design continued the contemporary motif, selecting furniture and interior finishes in a warm, monochromatic spectrum with plenty of texture. Combining these elements, Vesu wields a commanding presence — one that draws customers, welcoming them into an open, communal space that is simultaneously sophisticated, comfortable and warm.

设计师沿用当代的设计主题，选择多种材质的、统一的暖调家具和内部装饰物，巧妙融合。Vesu 餐厅有一种迷人的特点：引人入胜的气质、开放共享的氛围，兼顾精致、舒适与温馨。

M
W
W

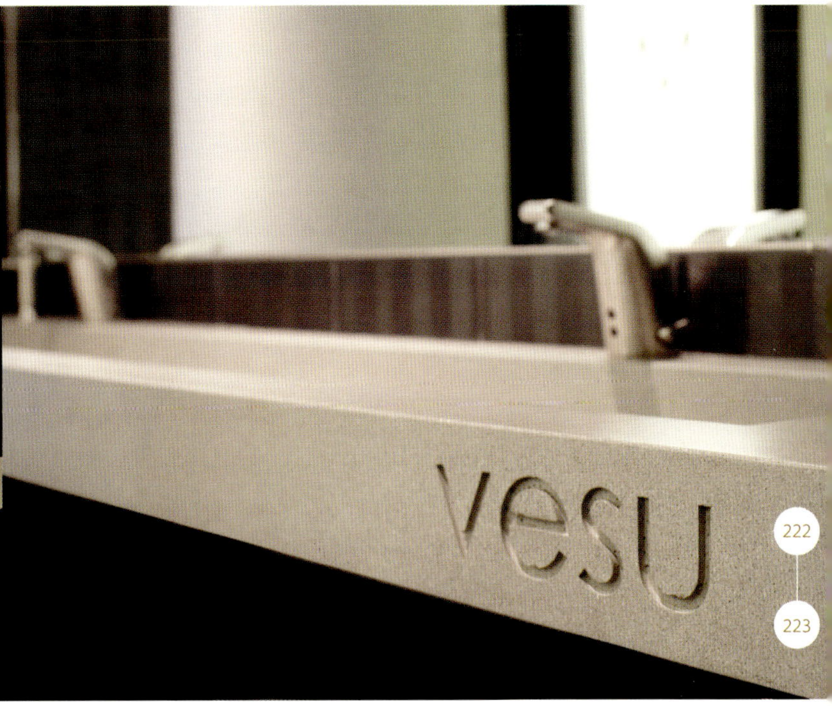
vesu

“吐司”餐厅

Project > 项目名称 > “吐司”餐厅

Architect > 设计公司 > Stanley Saitowitz/Natoma 建筑公司

Location > 项目地点 > 美国旧金山

Photographer > 摄影师 > Rien van Rijthoven

项目介绍 PROJECT DESCRIPTION

Toast restaurant is the second branch at the new Hamiltonplace shopping mall in Navato, north of San Francisco. It serves comfort food for breakfast, lunch and dinner. Comfortable environment and the selection of materials is its biggest characteristics.

餐厅位于旧金山北部，纳瓦多市哈米顿购物中心，这是它的第二家分店。这里供应早、中、晚三餐，舒适的环境和别具匠心的选材是该餐厅最大的特色。

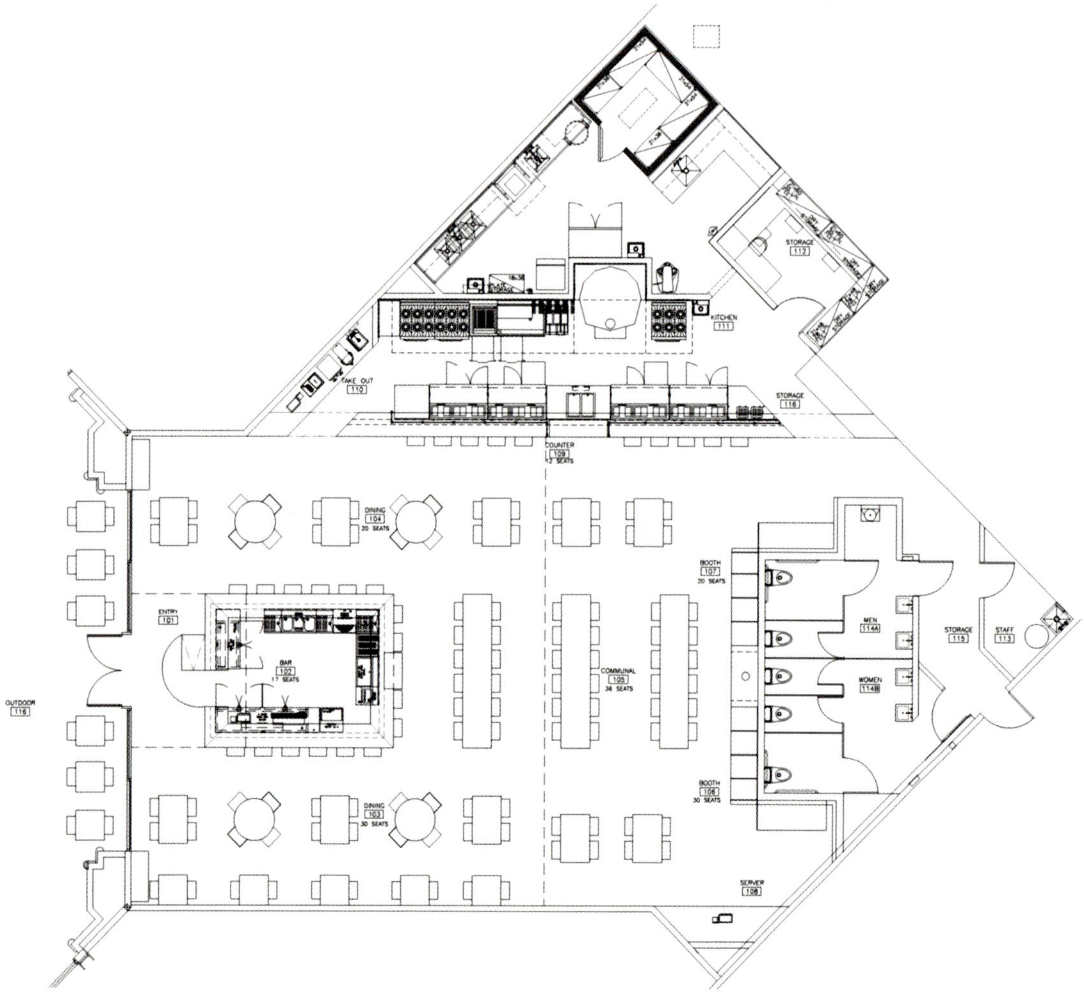

TOAST

The experience of entering Toast is like walking inside a loaf of bread, like swimming in sparkling Champaign…The yeast that creates this fizzy interior world is particle board, perforated with random shaped holes, which covers all surfaces. Walls, ceilings, furniture are all wrapped in this bubbling, bread like material.

步入"吐司"餐厅就像踏上一条软软的面包，在闪闪发光的香槟酒泳池中游泳。刨花板构造了一个布满"泡沫"的空间，盖满了空间的每一处表面，板上有各种形状的小洞。墙壁、天花板和家具都采用这种像面包一样的材料。

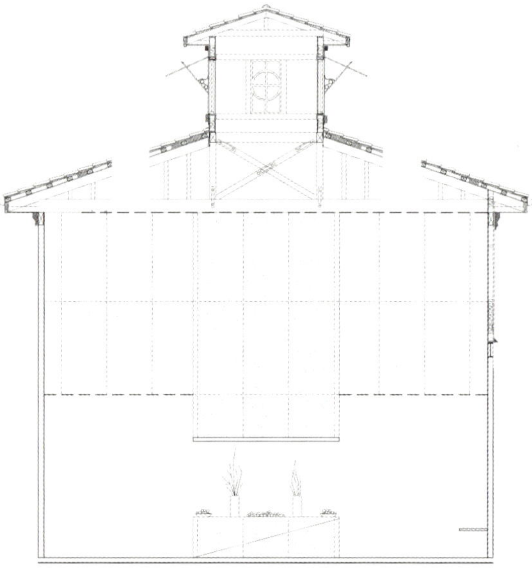

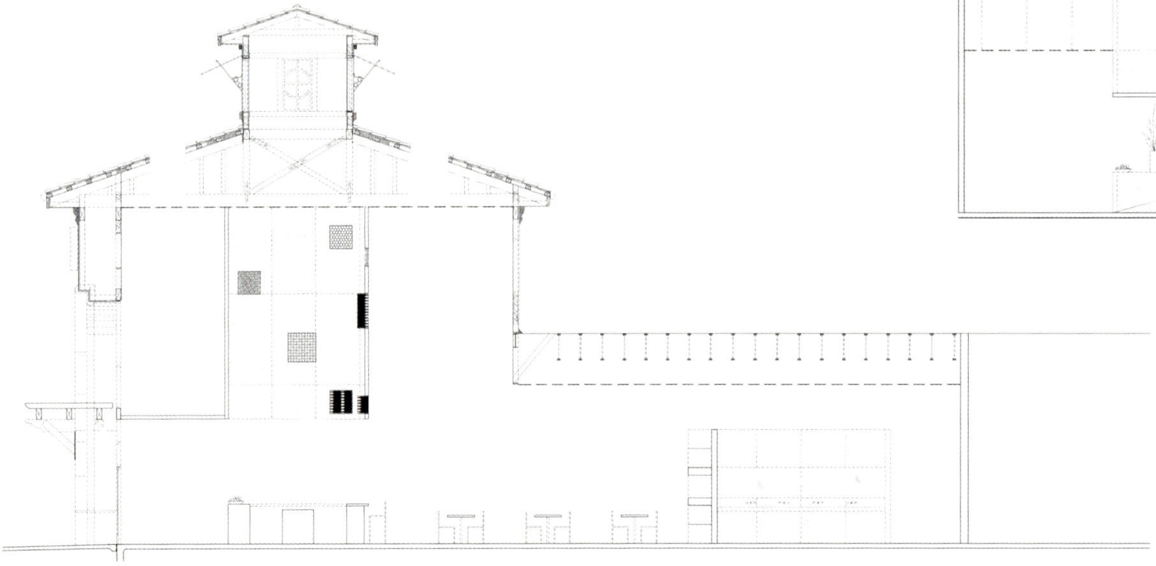

The low canopied entry transitions to a 40′ volumetric space with a hanging column suspended over the bar. This column has storage for glasses raised in the air. The bar mirrors this cubic cupola below, with seating on three sides. Behind the bar the ceiling drops to the main dining area. First are large communal tables, then smaller tables and booths.

低矮入口处有天棚，穿过这里就到达了室内，吧台上方悬挂着特制的筑体，上面可以用于储存玻璃器皿。吧台与筑体相互映衬，三面都设计了座位。吧台后方是主要用餐区。首先进入视线的是大的公共餐桌，其次是小一些的餐桌和双人餐位。

The open kitchen has bar seating and becomes a stage for the chefs and pizza makers. A fireplace with storage niches warms the space. Niches in the toasty walls are for displays of ingredients and wares. Behind, the toilets are contained in a floating box entered from the rear. The image is homey, warm as toast, making guests comfortable.

开放式厨房也安排了座位，厨师和披萨制作师是这里的主角。壁炉四周安装了可以储存货物的橱柜，给空间增添了亲切气息。这些"吐司墙"上的橱柜还可用于展示材料和商品。壁炉后面的一个可移动的盒子是卫生间，可以从餐厅后门进入。整个空间呈现出一种舒适、温暖的气氛，如同吐司面包一样温馨。

“第二个家”厨房式餐厅

Project > 项目名称 > “第二个家”厨房式餐厅

Architect > 设计公司 > Andre Kikoski 设计公司

Location > 项目地点 > 美国

Built area > 项目面积 > 465 m^2

Photographer > 摄影师 > Eric Laignel

项目介绍 PROJECT DESCRIPTION

The Owner of this Denver hotel restaurant asked the design team to conjure up the feeling of someone's Aspen or Telluride vacation home, instead of a stereotypical hotel restaurant. The Second Home creates the feeling of being at someone's house for dinner in a rich, warm and welcoming environment by drawing inspiration from the character of the Rocky Mountains and the people who live there.

这个位于丹佛宾馆的餐厅要求设计团队能做出像 Aspen 或 Telluride 度假小屋的感觉，而不是一个老套的宾馆餐厅。“第二个家”的方案意在营造一种在富裕、温暖和轻松的家庭环境中就餐的感觉，这个创意受到洛基山脉的别墅及当地居住特征的启发。

1 - VESTIBULE
2 - LOUNGE
3 -BAR
4 - FIREPIT
5 - OUTDOOR SEATING AREA
6 - MAIN DINING AREA
7 - SEMI PRIVATE DINING
8 - PRIVATE DINING
9 - KITCHEN/SERVERY
10 - KITCHEN
11 - MANAGER'S OFFICE
12 - WINE ADEGA

Second Home transforms what was once one large 5,000 SF venue into series of spaces that feel like the rooms of a home: the private dining room is a study, the main dining room is an open kitchen / family room, the bar a recreational room, the semi-private dining room a more formal dining room, and the courtyard a backyard patio with the Aspen trees and open flame.

该案将一个比 465 平方米的旧金山会场大一倍的场地变成了一系列像家的房间：私人餐厅好比书房，主餐厅是一间开放式的厨房和家庭活动室，酒吧为娱乐室，半私人餐厅是一间更正式的饭厅，院子中种满了白杨树并且有一处明火台。

The materials selected to underpin the authenticity of Second Home is a part of the Rocky Mountains. Rich textures like pony-skin upholstery and paneled walls, bark tile, and reclaimed wood contrast with stone, stainless steel, and glass to create an intriguing and thoughtful material palette. The sober geometry of the plan is complimented by unembellished dry-stack stone walls and rough, wood-plank ceilings. Design elements such as graffiti-printed upholstery on Arne Jacobsen Egg Chairs and artisan-quality 1950s Italian chandeliers accent the space, bringing a stroke of whimsy and animation.

"第二个家"的结构加固所选用的材料都来源于落基山脉。马皮软装材料、镶板墙、树皮纹瓷砖、与石材形成对比可回收木材、不锈钢、玻璃，他们考究的组合形成了迷人的搭配效果。未经修饰的干压石墙和粗木板天花板制造出干净的几何元素。设计师还善于利用带涂鸦的雅各布森蛋形椅罩和 20 世纪 50 年代风格的意大利手工枝形吊灯等元素，带来一种与众不同并且富有动感的视觉冲击。

Second Home effortlessly knits together the comforts and joys of a vacation home with the modern aesthetics and wit of a sophisticated restaurant. The space envelopes diners in warmth and hominess, and juxtaposes the austere and theatrical with architecture that grounds the restaurant in its Colorado setting. The result lends a distinct sense of place that draws back customers again and again. Second Home's charisma has made it a local favorite. It is a place people genuinely want to be a Second Home.

"第二个家"餐厅简单地将度假屋的休闲舒适与高级餐厅的现代美感和智能相结合，使来宾可以享受温馨舒适的用餐环境。建筑将朴素与戏剧色彩统一于科罗拉多风格之中，成为吸引顾客再次光顾的噱头。"第二个家"已经变成了当地的一大特色，成为一处人们发自内心地想把它看成第二个家的地方。

More than simply aesthetics, Second Home is born of Colorado. All materials, from the stones to woods, were all locally sourced, and local craftsmen played an integral role in the execution of the design.

除了美学考虑外，"第二个家"的方案还极具科罗拉多特色。所用的材料从石材到木材，均来源于本地，并且在设计执行过程中，本地手工匠人也担当重要的角色。

瑞士小屋湖餐厅

Project > 项目名称 > 瑞士小屋湖餐厅

Architect > 设计公司 > CCS 设计公司

Location > 项目地点 > 美国

Built area > 项目面积 > 930 m^2

Photographer > 摄影师 > Paul Dyer

项目介绍 PROJECT DESCRIPTION

In the city of Oakland locates the Lake Merritt, the city's historic urban lake and a famous landscape. CCS Architecture transformed the Mediterranean Revival style building into Lake Chalet Seafood Bar and Grill, a day and nighttime destination that has brought a surge of new energy to Lake Merritt.

米瑞特湖位于加利福尼亚州奥克兰市，是当地著名的景区，代表了该城市悠久的历史。CCS 设计公司将地中海复兴风格的建筑改造成了现在的瑞士小屋海鲜酒吧烧烤店，成为游客白天和夜间的好去处，为湖区带来了一股新能源。

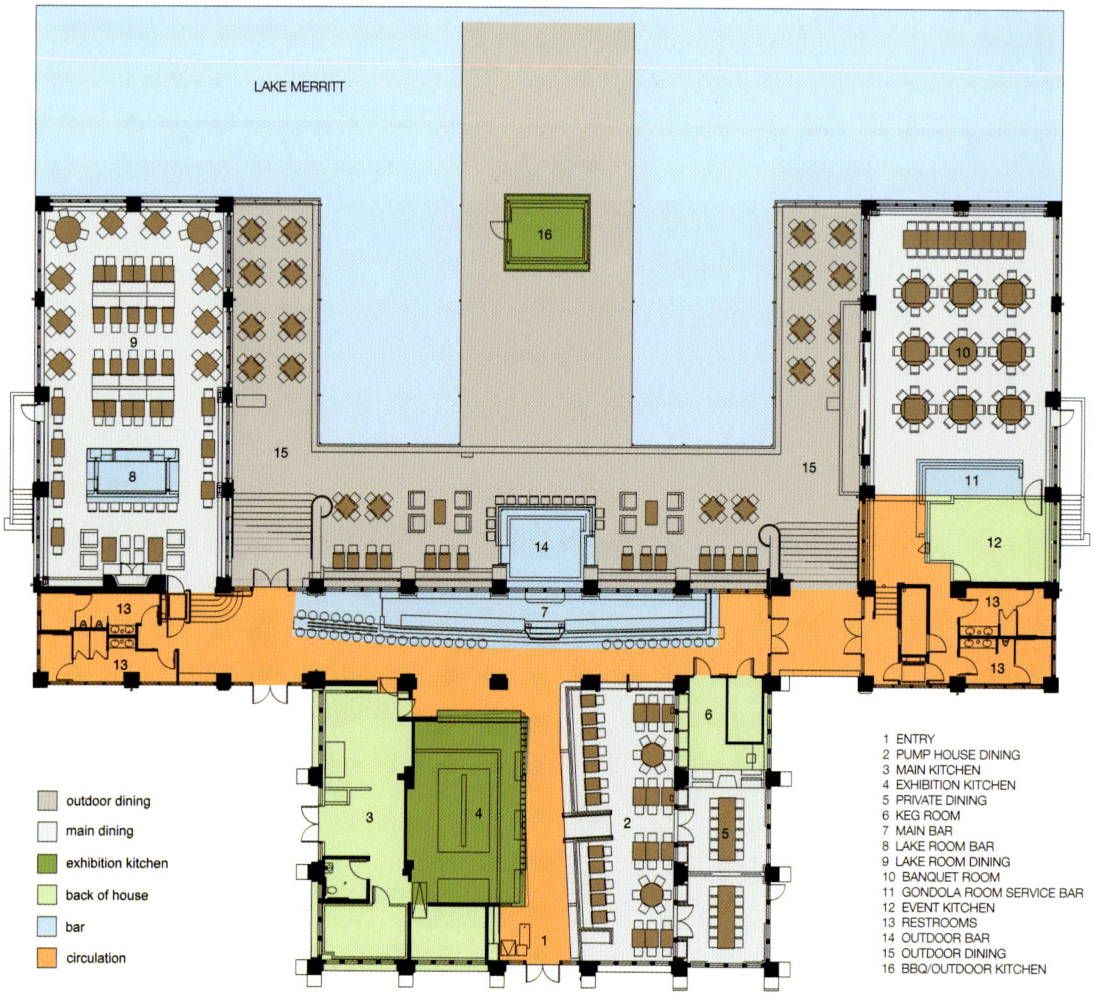

Lake Chalet. Laket Merritt, Oakland, CA

CCS ARCHITECTURE

LAKE CHALET
LAKE MERRITT BOATHOUSE

LAKE CHALET

With 325 indoor and 100 outdoor seats, the restaurant visually and physically embraces the waterside setting, with its walking and jogging path, gondola rides and "necklace of lights" dating to 1925. New architectural elements, materials and colors pay homage to the lake and the long history of boating at the site.

325 个室内座椅、100 个室外座椅，整个餐厅依湖而建。在湖堤上信步，或揽一叶小舟，或欣赏湖边的复古灯饰，恍如隔世。一些现代的建筑元素、材料、色彩为这汪湖水增添了色彩，仿佛诉说着米瑞特湖上泛舟的漫长历史。

On warm evenings, Lake Chalet's dock patio is packed with patrons enjoying cocktails with views of the lake and setting sun reflecting off the East Bay hills. Red umbrellas, striped awnings, and brightly colored fabrics create a festive ambiance.

夏日的傍晚，小屋的露台上拥着许多人，他们或品尝鸡尾酒、或观赏湖景或欣赏落日。红色的大伞、条纹的遮阳伞、艳丽的布匹显得欢快热烈。

LAKE CHALET

LAKE CHALET

Servizio Gondole
Gondola Tours

A casual dining area adjoins the Pump House Bar; the focal point here is the new, semi-open kitchen that CCS inserted into the historic structure. Inspired by boat-building methods, the kitchen's lacquered blue walls are shaped like the hull of a boat. An expressed structure of vertical mahogany ribs alternately reveals and obscures views of the cooking action. Openings are glazed with semi-transparent amber glass that suggests fishermen's net floats. The kitchen is a striking new piece of architecture within the expansive, historic building.

休息餐厅同露天酒吧毗邻，设计师增加了一个新的半开放式厨台，从造船获得灵感，整个厨台的外壁被漆上了蓝色，就像船外壁的保护色一样，直立的红木柱隐约地展现了里面厨师的动作，木柱之间的开口镶了一层半透明的琥珀色玻璃，好似捕鱼用的鱼网漂。厨台在整个宽阔而具有历史感的场所中显得非常现代、引人注目。

Two dining rooms occupy the adjoining wings that seem to hover above the lake; they are built atop open-air boat storage that is used by the gondolier and the Lake Merritt Boating Club. Dubbed the Lake Room and the Gondola Room, the building wings have large windows on three sides that provide dramatic views as well as natural ventilation. New mahogany ceilings and restored mahogany windows complement the original metal truss ceilings.

餐厅的两边各有两间房，处在翼的位置，环抱湖面，楼下为船夫和泛舟俱乐部的露天的泊船地，一个叫"湖景房"，一个叫"船夫房"，三面都有巨大的窗户，可以看到湖景而且通风良好。红木天花板和窗框修补了原来的金属桁上的天花板。

The Lake Room is a romantic, upscale dining room with movable, freestanding banquettes, amber glass orbs, its own bar and an elevated wine bottle display. The Gondola Room is a special event space for 100 people, outfitted with hardwood floors, audiovisual capabilities and an adjacent catering kitchen. Its private balcony has sliding mahogany doors that open completely to the lake and dock below.

"湖景房"是一个浪漫而精致的餐厅，里面包括人行道、琥珀色的玻璃球灯、独立的吧台、高架酒具柜。"船夫房"是一间可容纳 100 名客人的特定空间，硬木地板，配有多媒体设备，旁边有一间厨房。穿过红木滑动门可到达独立的私人露台，开放式的设计使这里与湖景、码头融为一体。

25 卢克斯餐厅

Project > 项目名称 > 25 卢克斯餐厅

Architect > 设计公司 > CCS 设计公司

Location > 项目地点 > 美国

Built area > 项目面积 > 911 m^2

Photographer > 摄影师 > Paul Dyer

项目介绍 PROJECT DESCRIPTION

The 265-seat new American restaurant and bar is an unexpected gem in the urban fabric. CCS Architecture crafted the two-level space, weaving graceful forms and sophisticated materials through the massive, historic, warehouse structure. The interior emphasizes a counterpoint between the new palette of polished stainless steel, glass, white plaster, leather, mirror, faux fur, and slate and the existing elements of brick, concrete and rough-sawn timber.

这个 265 个座位的新式美国餐吧非常受欢迎。该案被建造成两层空间，利用优雅的流线外形和复杂精致的材料改造了这个具有历史意义的大仓储式结构。室内装饰突出不锈钢、玻璃、白石膏、皮革、镜面、人造毛、石板和砖、混凝土和粗锯木材间的色彩对比。

LOWER LEVEL FLOOR PLAN

1 BACK STAIR FROM ENTRY
2 BAR
3 MAKE OUT ROOMS
4 COMMUNAL DRINKING TABLE
5 RESTROOMS
6 LOUNGE SEATING
7 FIRE ORB
8 PRIVATE DINING
9 MAIN STAIR TO DINING ROOM
10 ELEVATOR

lounge
office/back of house
bar
private dining

OGDEN PACKING & PROVISION CO
25

TWENTY
FIVE
LUSK
25

OGDEN PACKING &
25

The architects created a large, glass entrance, cutting into the existing building exterior; the canopy bends up at its leading edge to become the restaurant's signage. Windows were enlarged and added along the façade to animate the interior with natural light and allow views. Inside, a large wedge from the upper floor makes an open connection between the lower level lounge and the dining room upstairs. Entering the restaurant, guests take in simultaneous views of both.

建筑师将已有的建筑外墙打断，建成一个大的玻璃门入口，将顶棚前端折弯用作餐馆招牌；扩大餐馆正面的窗户面积和数量，使餐馆内部因自然光线的照射和开阔的视野而生机盎然。将底楼休闲区与楼上饭厅之间设计成开放式相连，连接处形成一大片楔形空间。客人一踏入餐馆，就能同时将两层楼一览无余。

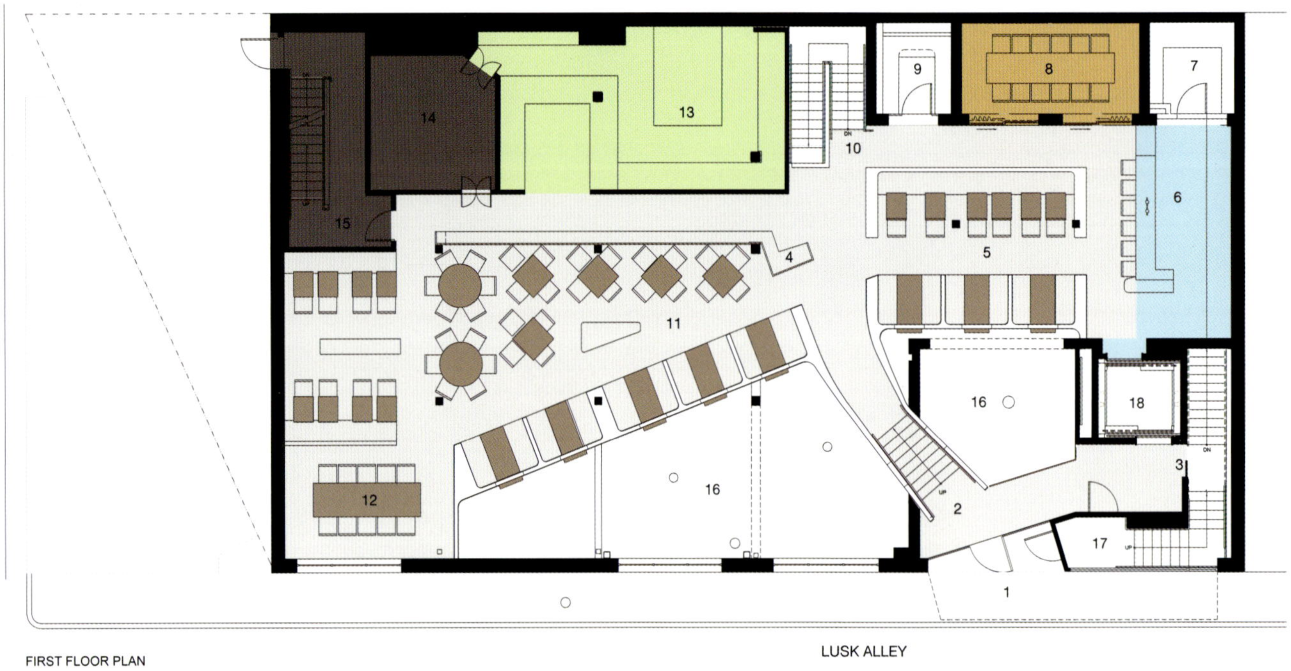

FIRST FLOOR PLAN

1 MAIN ENTRY
2 STAIRS TO MAIN DINING
3 BACK STAIRS TO LOUNGE
4 HOST
5 BAR SEATING
6 BAR
7 LIQUOR STORAGE
8 PRIVATE DINING
9 WINE ROOM
10 MAIN STAIRS TO LOUNGE
11 MAIN DINING
12 COMMUNAL TABLE
13 GLASS-ENCLOSED KITCHEN
14 WARE WASH
15 EXIT STAIRS
16 OPEN TO LOUNGE BELOW
17 STAIRS TO OFFICE
18 ELEVATOR

main dining + entry
back of house
kitchen
bar
private dining

25 Lusk. San Francisco, CA

CCS ARCHITECTURE

Behind the large bar, the former Ogden Packing and Provision smoking rooms have been converted into intimate lounge areas. These semi-private, brick and concrete chambers are appointed with sumptuous sofas. The lower level features a 40-seat private dining room as well as a glass-enclosed wine room within the former freight elevator shaft.

吧台后面原来的吸烟区被改成私人休闲区。在半开放的、砖混结构的私房内安放着奢华的沙发。底层共有 40 座的私人餐厅和由货运电梯改建成由玻璃围成的酒室。

In the lower level lounge, seating zones extend the length of the space, each with a suspended, stainless steel fire orb. The orbs act as a focal point for each seating area, much like camp fires, and their reflective flues extend up through the restaurant's open spaces to the ceiling 20 feet above.

在底楼休闲区，座位区域占满了整个空间，每个座位区域悬挂着一只不锈钢炉。它是每个座位区的焦点，作用与营火类似，其烟道从餐馆的开放空间延伸到天花板上方6米处。

The dining room is on the second floor, up a half-flight of stairs from the entry. The kitchen is a highlight on this level; a modern envelope of clear and black glass permits views of the chef action and reflects the activity of the dining room. A strategic mix of tables, banquettes, and booths provides seating for 120. Pullman-style booths are built into the angled, low plaster wall that borders the cut-away, and cantilevered tables, made from richly patterned Macassar ebony, piercing the wall. Lighting reveals the original Douglas fir ceiling and creates a warm glow.

餐厅在二楼，处在梯级一半的位置。厨房是这一层的亮点，厨房使用黑色透明玻璃，外面的人能看到大厨在里面的动作，从厨房里面也能看到外面餐厅的情况。餐厅内的餐椅、长椅、雅座全部加起来共 120 座。铂尔曼风格雅座由插入低矮石膏墙内的望加锡乌木纹悬臂桌分隔开。灯光照射到由道格拉斯冷杉原木制成的天花板上，营造出一种温暖的氛围。

杰米的意大利餐厅

Project > 项目名称 > 杰米的意大利餐厅

Architect > 设计公司 > Blacksheep 设计公司

Location > 项目地点 > 英国伦敦

Built area > 项目面积 > 710 m^2

Photographer > 摄影师 > Gareth Gardner

项目介绍 PROJECT DESCRIPTION

Blacksheep has designed the latest "Jamie's Italian" restaurant for Jamie Oliver at the Westfield London shopping centre. The "Jamie's Italian" restaurants are Jamie Oliver's first completely independent restaurant venture and aim to bring "what's best about casual dining" to the high street.

Blacksheep 设计公司为杰米奥利弗设计了一家意大利餐厅，它位于伦敦西田购物中心，这是杰米奥利佛第一家独立开办的餐厅，为的是给主商业区带来一流的休闲餐饮理念。

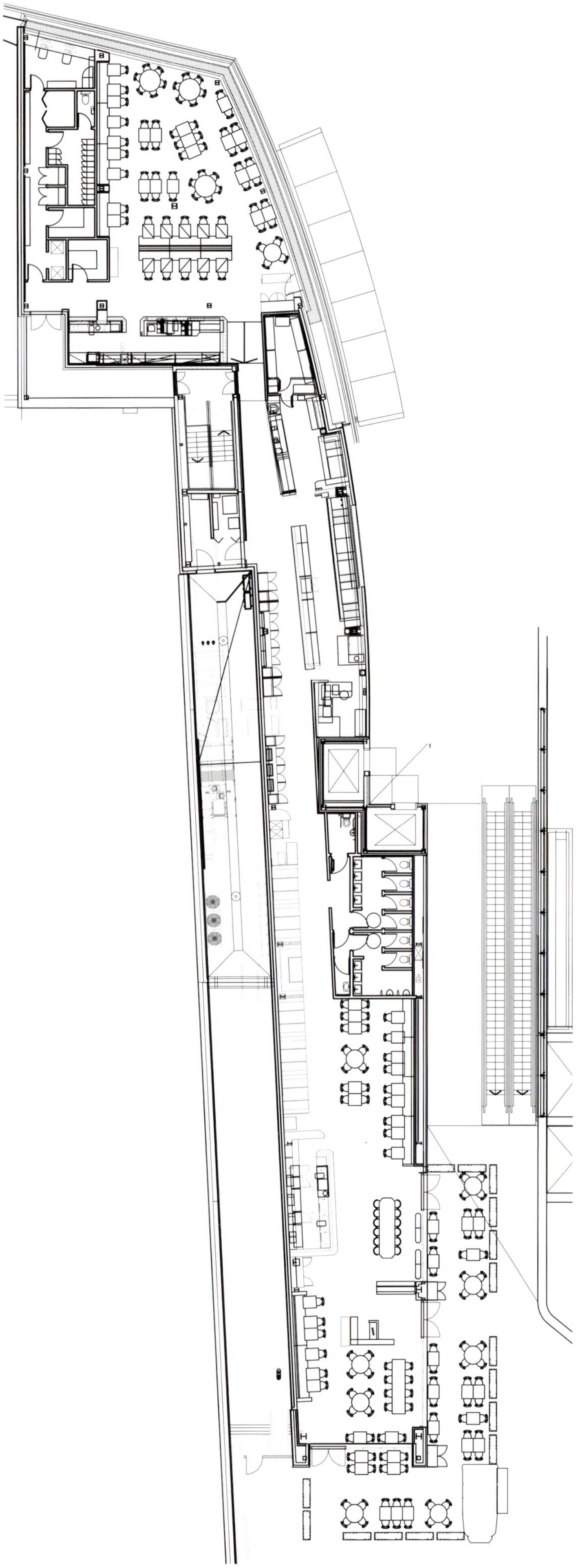

The restaurant exterior has a great L-shaped space (approx 100 sq m in total) on Westfield's Southern Terrace, which offers al fresco dining beneath a canopy, suitable for year-round usage, with all restaurant doors and full-height windows able to be opened out onto the space in warm weather. The new, branded ice cream van sits at the corner of the L-shape. The external area is bordered and demarcated by planters made from reclaimed scaffold boards, which are full of herbs, creating both a great smell and a sense of anticipation for the food to come.

餐厅外部是约 100 平方米的直角形区域，位于西田南广场，可在户外顶棚下就餐，而且一年四季都可以使用。天气晴朗时，所有的门和落地窗都开着，客人可以享受阳光。新添置的冰激凌厢式车位于这个区域的拐角处。再生材料做的脚手架上摆满植物，将外部区域分为多个部分，植物的清香提高了客人的食欲。

Overall, the interior has a pared-down industrial feel, with exposed gantries, steelwork beams and columns, dark grey paintwork and lots of timber. It is spacious, warm and accessible but doesn't pretend to be what it isn't. Flooring is a mix of recycled, engineered timber and Black Mountain river slate (used in three different sizes in random patterns).

暴露在外的吊架、钢制横梁、圆柱以及深灰色的油漆和木材给内部空间添加了些许工业感觉。宽敞、亲切的环境是无法复制的。地板的铺装使用的是再生复合木板和黑山板岩，根据不同的形状设计了三种不同的尺寸。

LEVONI

AMAZING ROLLED AND OVEN
BAKED CORNISH SARDINE FILLETS
FILLED WITH HERBY BREADCRUMBS,
PINENUTS + RAISINS SERVED
WITH A DELICIOUS BLACK OLIVE
TAPENADE. £12.95

The 610 sq m interior is very long, with a particularly narrow central area, which wasn't ideal for seating. From the main restaurant space ("Piazza") to a central "Market Place" area, without seating, which made a virtue of the narrow central space and where customers can see views of the kitchens and fresh pasta making or can linger over a bigger retail area than ever before, en route to the restaurant's rear dining space. The Back Room is a special and more intimate area with lower ceilings and a more slightly "bling" treatment.

约 610 平方米的内部狭长空间，中心区域最窄，不利于座位摆放。Blacksheep 公司设计了一个合适的方案，从餐厅的主要就餐区到中心区域都没有摆放座位，客人可通过中心区域直接看到厨房，观看到意大利通心粉的制作过程，或者观察宽阔的零售区，直到后面的用餐区。被称为黑房间的包间是一个特别设计的私密空间，天花板较低，装饰较为豪华。

WC

Wall treatments include re-claimed ceramic bricks in the main restaurant, feature metal panelling around the toilets area and bespoke timber and metal panels in the Back Room.

主餐厅的墙面使用再生陶粒砖，卫生间使用的是特质金属板，“黑房间”则使用定制的木板和金属板。

Whwm 餐厅

Project > 项目名称 > Whwm 餐厅
Architect > 设计公司 > The Metrics 设计公司 （Elle Kunnos de Voss）
Location > 项目地点 > 美国纽约

项目介绍 PROJECT DESCRIPTION

The overall "work in progress" concept for the space is designed to reflect the changing and experimental nature of the project. With our actual architectural drawings projected onto the surfaces of the space in scale 1:1, the guests are invited into the design process. To serve as a backdrop for the monthly changes we inverted our drawings to give the functionality of a theater black box. The ceiling is covered with a 12" grid of hooks to keep the space flexible and to be able to easily reconfigure the lighting for each Movement.

整体设计理念反映了项目的变化和设计的特点。设计师将一些建筑图纸上的标记以 1:1 大小直接刻画在地面、墙面和屋顶上，客人们也能参与到其中。由于每个月都要进行主题的更换，设计的图纸模仿了"戏剧黑盒子"。天花板上挂着钩子，有助于空间使用的灵活性，每次灯具的改动都得益于这些钩子。

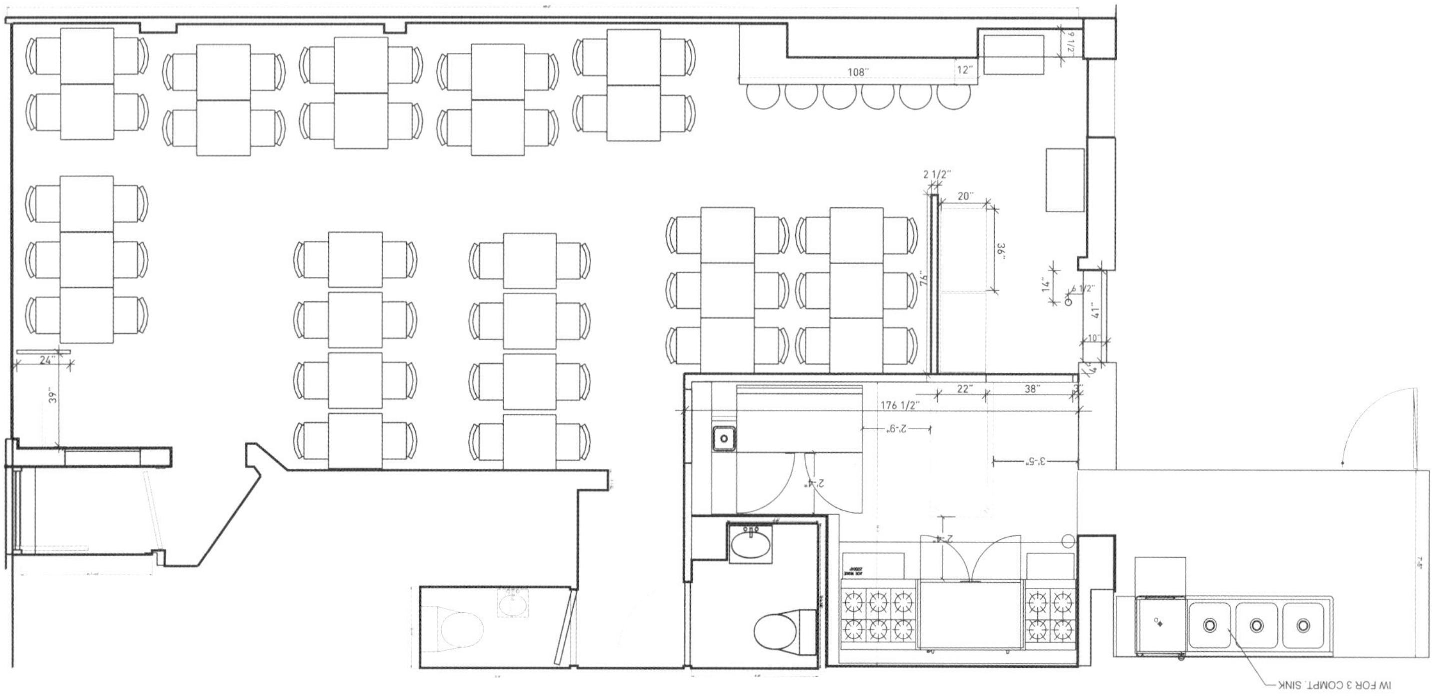

01
A 4.0
NORTH WALL ELEVATION
SCALE 1:1

PT-01

EXIT

01
A 4.0
NORTH WALL ELEVATION
SCALE 1:1

Movement 1 was inspired by a Nordic palette where the space was designed as an achromatic landscape of deconstructed volumes and fixtures, using a paired down aesthetic. White lines defined volumes within the space, describing archetypal house and ladder shapes in a distorted perspective. The deconstructed chandeliers take their cue from a classic chandelier with cut cardboard prisms and large globe light bulbs. Fabric installation with one big stroke of hot pink fading into deep purple makes the space more intimate.

主题一受北欧风格影响，整个空间由黑白色调的解构体组成，符合对称审美观。白色线条勾画空间的形状，房子的原型和梯形装置达到扭曲的透视效果。屋顶的框架为空间打造了更多餐桌，不同角度搭建的梯子作为房间隔板使用。解构的枝形吊灯搭配纸板做成的棱镜以及圆形大灯泡，显得格外复古。空间内使用的布料以桃红色和深紫色为主，使空间在视觉上更亲切。

FAYETTE STREET
25
NORTH

PT-01

With the forest where the wild things are theme of Movement 2, the interior design concept for Movement 2 is taking on a play with scale. The space is defined by an installation of over-sized pine needles that create a movement across the ceiling. Stretching to the floor in some areas the pine needles act as room dividers. Throughout the space little moments unveil the fantastical forest theme such as two moss-laden swings with miniature landscapes of plants and birds, bird houses nestled in the pine needles and various animal tracks on the floor, walls and selected tables.

主题二是户外森林，比例上与戏剧布景相似。空间内布置了一个超大尺寸的松针装置，穿过屋顶。一些区域的松针延伸到了地板，起到了分隔空间的作用。空间内布满了梦幻般的森林主题，例如两架用苔藓做成的秋千，周围摆着微缩盆景和鸟类模型，鸟巢安在松针里，地板上、墙上和桌子上到处是动物的“足印”。

Movement 3 takes on a spring garden party theme. The interiors draw a few significant elements and from the Renoir painting to re-create an intimate, communal experience reminiscent of 19th century time and place. Creating a play between interior and exterior, elements such as branch-like light fixtures and iconic Still Life Wall sconces nod both to nature and the genre of painting.

主题三布置成了春季游园会的场景。内部的陈设大多是一些具有实际意义的元素，设计师从雷诺瓦的画中受到启发，重塑了一个亲切的公共空间，令人联想到 19 世纪的时光。餐桌的摆设像在温泉中一般，重塑了画中的意境。内外统一的元素如树枝一样的灯具、烛台等陈设显得自然又符合整体的艺术风格。

01 NORTH WALL ELEVATION
A 4.0 SCALE 1:1

Inspired by Jazz, Movement 4 explores the tension between improvisation and tradition, as well as the unique rhythm and architecture that define this musical form. This spatial rhythm visually and spatially communicates the rhythm and contrast of Jazz. The entryway welcomes guests with a taste of what's to come with a collage of patterns and imagery that mix old photographs from New Orleans with abstract patterns that visually represent the tempo of jazz. Adding another "note" of reference, all the light fixtures are inspired by jazz line up instruments. Made from basic geometric shapes in bright colors, they are configured around the room at different heights, adding to the visual symphony.

受爵士乐的启发，主题四在即兴设计与传统设计中寻求平衡。空间的风格在视觉上和空间上与爵士乐有共通处，也有对比性。入口处的拼贴画将新奥尔良旧照片与抽象图案融为一体，视觉上象征了爵士乐的拍子。灯具被设计成了爵士乐器的形状，均为亮色调的几何模型，布置在不同的高度上，奏响一曲视觉交响乐。

SZAFRAN
SUSAN MILL
SARAH ULLMAN
AKI NAITO
JED GIDALY
EATER.COM
STEVE CLARK
FRED WILSON
YAYI BAZO
KENDEL RATLEY
MICHELLE WAGNER
ASTRID
KATE WOOD
RIZI
CARMEN MELLO
MASON ADAMS
EMILY
DUNCAN WINTER
LEN SMALL
SILVER
HANLON
TURKELL
KATE HOEHLEIN
ALBERT WENGER
JON FINE
SUSANNAH SHIPMAN
JOEL ZIMMER
KATE FARBER
KYLE CAMERON
CORY FORSYTH
DAVID BERKOWITZ
SCOTT KESTER
MATTHEW WARD
CAMERON SCHILLER
CARRIE GABODA
JAMES SHALOM
CHRISTOPHER AIDUN
CIARA BURNHAM
ILENE DEUTSCH
CHRISTIAN SUMNER
DENA ROSENBERG
DAVID NISINSON
VICTOR YEN
TAMAR SIMON
LUCE JORDAN
BALLERINA
ERIC WOLFF
ROBIN REED
MICHAEL KARNJANAPRAKORN
KIT HAVES
AMAYA CAPELLAN
WILL SCHENK
GEORGE RADWAN
JOEL COHEN
DAVE DICAMILLO
CARLYE ADLER
BATHROOM 2

TITLE: EAST WALL ELEVATION
DRAWING:
A-201
DATE:
SCALE:

PT-01

EXIT
A-201
PT-01

PT-01

01
A 4.0
NOR
SCAL

EL Cardenal 餐厅

Project > 项目名称 > EL Cardenal 餐厅

Architect > 设计公司 > Pascal Arquitectos 设计公司

Location > 项目地点 > 墨西哥

Built area > 项目面积 > 700 m^2

Photographer > 摄影师 > Juan Marcos Castañeda

项目介绍 PROJECT DESCRIPTION

"El Cardenal Restaurant" is located in the ground floor of a modern building in the midst of a trendy commercial area of Mexico City. The contemporary design of the premises is in direct relationship with the sorroundings rather than with the traditional mexican cuisine, allowing it to be the centre of attention and leaving everything else functioning only as background.

EL Cardenal 餐厅位于墨西哥城繁华商业区中心一处现代建筑的底层。设计与周围的现代环境协调，而非传统的墨西哥食品店，使其成为目光的焦点，而周围的一切只能成为陪衬的背景。

The bar area is the principal node from which the complete distribution and the restaurant concept revolves, at the same time is the host zone that welcomes the visitors with a magnificent nava wood and black greased steel bar that gives access to the main foyer.

In the main dining room, a black Oaxaca's clay cabinet will display temporary expositions of selected pieces of the Museo de Arte Regional – MAP - (Regional Folk Art Museum); a loose ceiling panels covered with scantex acts like an acoustic barrier and the gaps between these panels and the black background helps to hide air conditioning outlets, speakers, sprinklers and the rest of the mechanical elements. Also from the dining room you can access an indoor yard that is in complete harmony with whole decoration concept.

The integration of the kitchen area was accomplished by using amber glass working as filter and extending the black ceiling from the main foyer towards it, creating a contuinity feeling between this two spaces. The floors were covered with natural and polished pórfido de San Luis stone; nava wood and a steel plate was utilized for the baseboard.

吧台区为主要节点，整个餐厅布局及理念皆围绕其设计。吧台区同时又是待客区，欢迎四方来客。由豪华纳瓦木质和黑色油彩钢材料制造的吧台正对大堂。

主餐厅内一个黑色的瓦哈卡黏土陈列柜展示了当代民间艺术博物馆的作品。疏松的顶棚镶板会起到隔音的效果，而且这些镶板的间隙以及黑色的背景可以帮助隐藏空调排气口、扬声器、洒水器以及其他电气设备。从餐厅区可进入一个室内天井，天井的设计与整体的装饰理念也完全相符。

厨房区的整合是通过琥珀玻璃实现的。琥珀玻璃可作为滤光器，黑色顶棚从大堂一直连接至厨房，在这两个空间之间便营造出一种衔接感。地板铺以天然抛光的圣路易斯斑岩石，护壁板使用了纳瓦木及钢板。

AB 咖啡店

Project > 项目名称 > AB 咖啡店

Architect > 设计公司 > Puresang 设计公司

Location > 项目地点 > 比利时

Photographer > 摄影师 > Frank Gielen

项目介绍 PROJECT DESCRIPTION

The objective for the new AB café and concert bar was to create an environment where the Stella Artois brand is fully integrated. The world of AB and Stella flow into each other and amplify each other in a very natural way.

这家新 AB 咖啡音乐吧的老板与啤酒品牌 Stella Artois 紧密合作，缔造出了一个和谐自然的设计理念。

1 PLAN CAFE

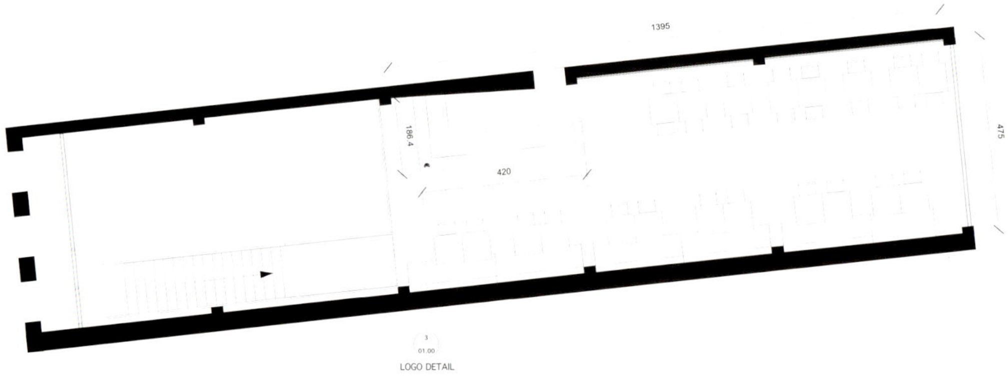

2 PLAN RESTO

Not by using screaming logos, but by working with subtle and creative allusions and an original contemporary design, the new concept that is used in AB, is called "glamour rock". Big graphics and a wall with "memorabilia display cases" let the rock level lead. A big chandelier of acoustic guitars gives the space a big touch of glamour.

店铺中没有非常显眼的 LOGO，但经过巧妙的创意和典型现代元素的设计，整个空间呈现出一种"华丽摇滚"风。巨大的装饰图案和"摇滚纪念品展柜"创造出了这种感觉，一个枝形吊灯上悬挂着巨大的木吉他，让空间显得极具吸引力。

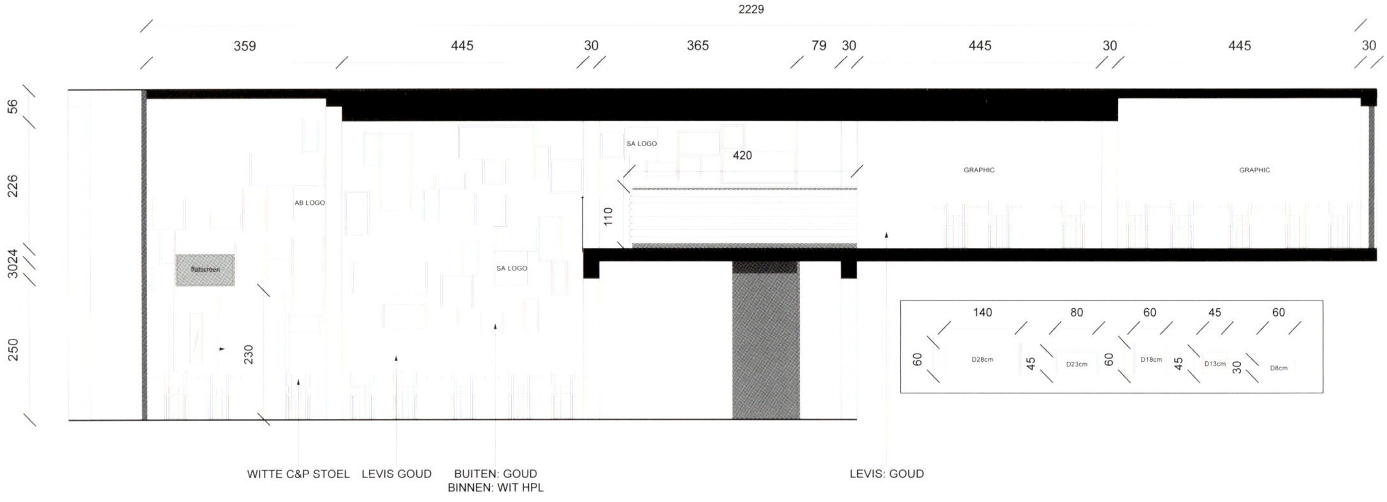

1 ELEVATION

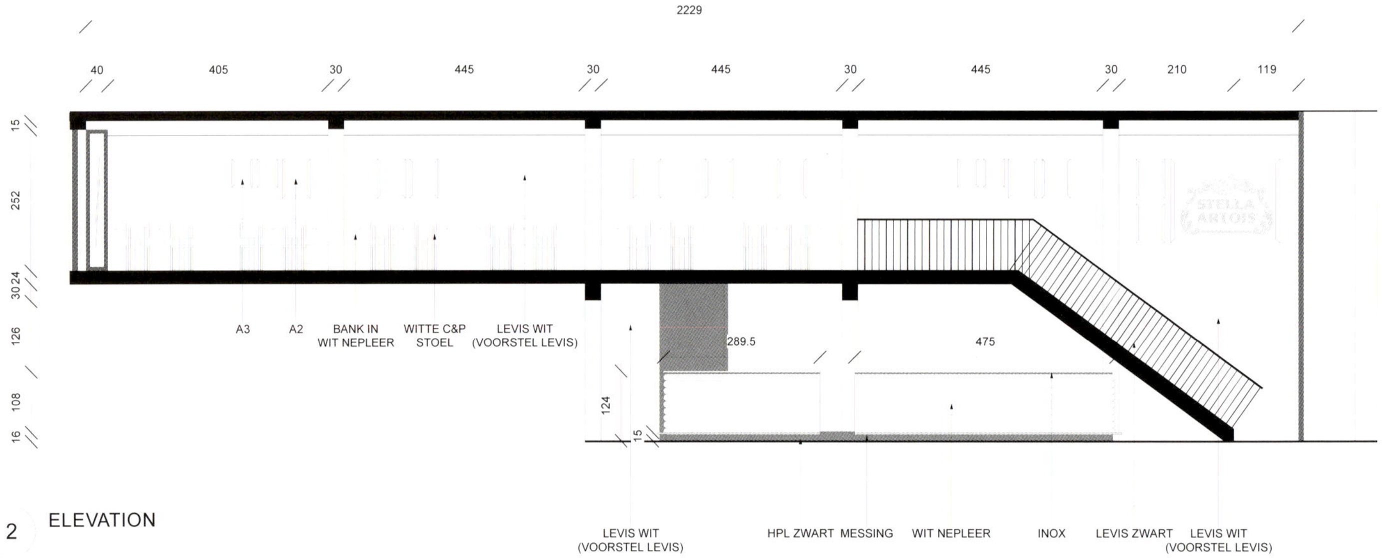

2 ELEVATION

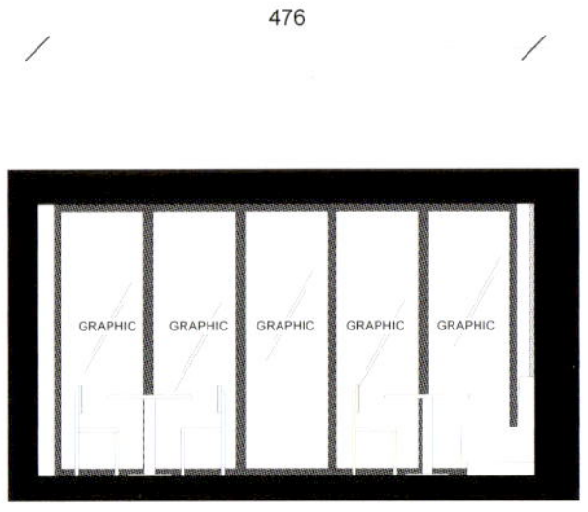

3 ELEVATION

euwen 1366
BOSE

STELLA ARTOIS

The golden graphics, in which Stella Artois logos are integrated in a subtle way, start at the back of the bar and continue over the window. The main colors are white and gold, but together with the colorful posters and memorabilia, the result looks balanced without being cold and damp.

从吧台后面一直延伸到窗户的 Stella Artois 金色 LOGO 显得更加精巧。空间主色调为白色和金色，在五彩缤纷的海报和纪念品的衬托下，不会显得过冷或呆板。

The backbar is a composition of high gloss blocks, creating a layered texture which integrates the backbar equipment. Flatscreen televisions, in several sizes, are mounted in this three-dimensional wall showing visuals of sponsors or extracts of AB TV.

后壁是一块高光板，很有层次，上面安装了一些设备。立体墙面上有几个尺寸、大小不一的宽屏电视机，播放着赞助商的广告和 AB 电视台的节目。

The AB restaurant upstairs has been put through the same transformation as the AB café. The original bar is covered in the same way as the bar downstairs. White leather couches give enough space to sit, and that makes reorganizing the tables for bigger groups much easier.

楼上的 AB 餐厅和 AB 咖啡吧有异曲同工之妙。餐厅采用和楼下一样的经典吧台，白色皮质长沙发给顾客构建了更多的休息空间，也便于更多的顾客拼桌使用。

808 休闲吧

Project > 项目名称 > 808 休闲吧

Architect > 设计公司 > plajer & franz 设计公司

Location > 项目地点 > 德国

Built area > 项目面积 > 250 m^2

项目介绍 PROJECT DESCRIPTION

The term "lounge" was established in the sixties in America and was the given theme for the project "barlounge 808" in Berlin mitte. Situated on a corner, the 250 square meter space is wedge-shaped. Floor to ceiling windows run a length of 40 meters, fronting two streets and meeting in a glass curve at the street corner. The bar lounge 808 is divided up into two areas, with a cafe/bar in the front serving food and drinks from midday, and a cocktail lounge. Design elements like the aquarium, light coves and custom furniture determine the relaxed atmosphere in this space. Speakers and air-conditioning are invisibly integrated in the wooden wall paneling.

"休闲吧"一词诞生于美国 20 世纪 60 年代背景下，位于德国柏林米特市的 808 休闲吧就是围绕此主题设计的。这个 250 平方米的楔形空间坐落在两条大街的相交处，立面安装了 40 米高的落地窗，好似在街角上安装了一面流线型的玻璃墙。整个空间分为两个区域：咖啡吧、酒吧设在正面，从午间便可提供饮食；另外还有一个鸡尾酒休闲区。空间中加入了许多设计元素，如金鱼缸、灯槽和定制家具，用于调节气氛，扬声器和空调隐藏在木墙板背后。

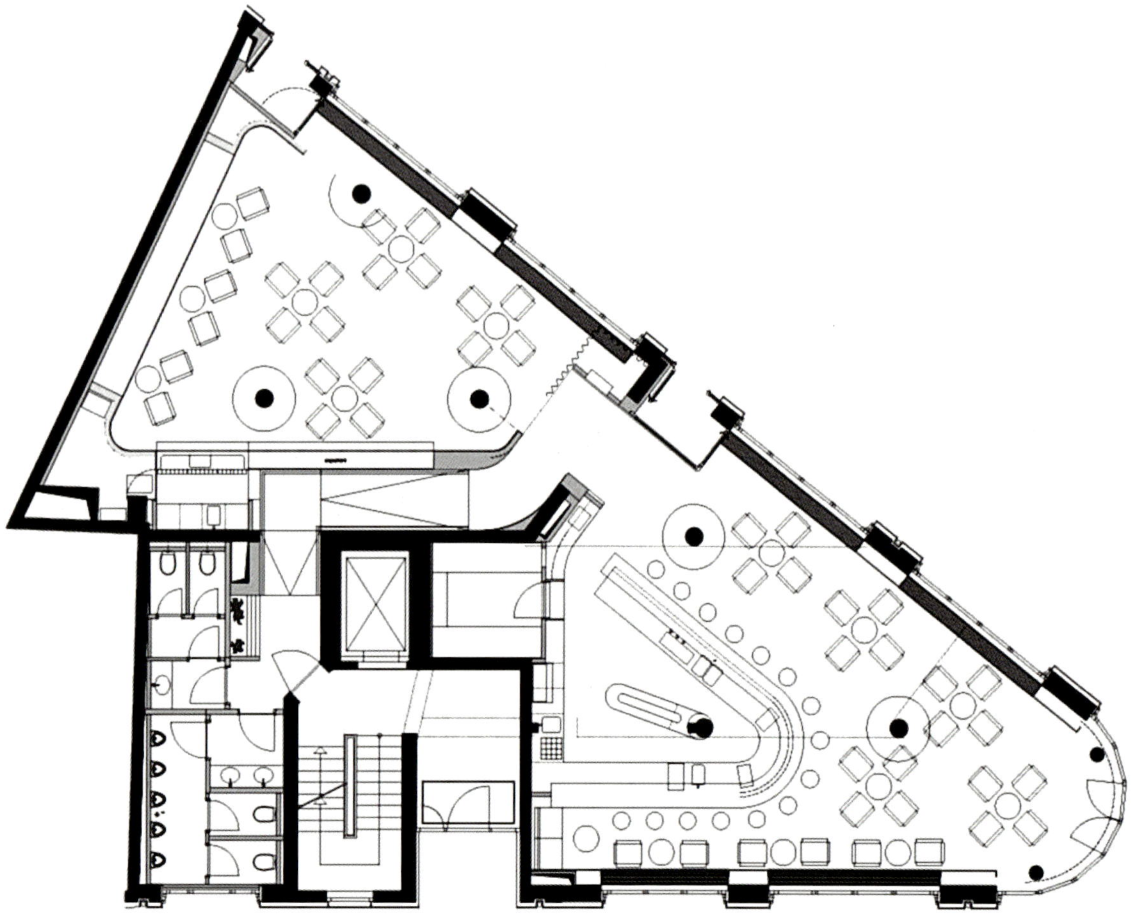

The chestnut room is far more louche affair. Diffuse lighting creates warmth and the wood and metallic hues reflect and intensify the effect. As in the café-bar, the metal framework of the furniture is powder-coated, this time in gold. The unusual "bubble" table tops are made from brass discs welded together, with a clear finish applied to prevent oxidization. A 2.5 tonne aquarium, measuring almost 5 metres (16 feet) in length, provides a relaxing focal point. Encased in the wood paneling it is a visual continuation of the small bar hatch. The illuminated tank also affords guests passing through the corridor to the restrooms a watery glimpse of the space.

棕栗色休闲吧里的气氛显得更奢华，散射光制造出温暖的感觉，木料和金属色调强化了这种感觉。咖啡吧的家具上使用了粉末涂层金属框架，呈现出金色。与众不同的"泡泡"圆桌表面用铜质的圆片焊接在一起制成，经过抗氧化处理，光洁如新。镶嵌在木墙板上的一个长约 5 米、重 2.5 公斤的大鱼缸增加了空间的休闲感，其与墙上的壁龛在视觉上相互呼应。经过洗手间走廊的客人们可以通过这个透光的鱼缸瞥见室内。

新月日式餐厅

Project > 项目名称 > 新月日式餐厅

Architect > 设计公司 > The Metrics 设计公司 （Elle Kunnos de Voss）

Location > 项目地点 > 美国纽约

Photographer > 摄影师 > Kazuhisa Niizawa

项目介绍 PROJECT DESCRIPTION

This Japanese Dining Spaces ambience is provided through the use of both beautiful lighting and shadow arrangements. An old Japanese Poem marks the frosted glass walls with a considerable amount of design consideration in featuring glass and bamboo in front of this.

华丽的光与影很好地烘托出这家日式餐厅的氛围。一首古诗刻在磨砂玻璃墙面上，墙前面的玻璃和竹子显示出设计师细密的构思。

The poetry was considered to reflect the balance of further beautifying the Japanese food with light. This offered lighting only in areas where it is really felt necessary while the diners need not be concerned with neighboring unknown customers.

光透过古诗词将日本食物照射得更加漂亮。灯饰不多不少，运用得恰到好处，在这样的光影下，食客们也不必担心受到周围陌生人的干扰。

The "Magic Box" Concept is the designer's original design concept that is now copied all throughout Japan. This box featured in the toilets, although only 20 cm with marguerite flowers lining the bottom, gives the impression that the room extends over several metres.

该设计师原创了现在日本非常流行的“魔盒”概念。餐厅的卫生间的底部装置了这种“魔盒”，邹菊成排地开放，将原本只有 20 厘米的底部从视觉上拓展成了几米。

Strip light installed on the stairs of the hallway are easy access from top to bottom. The lights are in the same style with the rear glass lamp on the side wall.

Scattering of light on the walls, private rooms and dining area creates a mysterious and elegant atmosphere. Wooden tables and chairs, bamboo curtain, tatami added a Japanese-style simplicity and freshness.

When wandering in each private room, you will feel it is in psychedelic lane. Before the Japanese cuisine you can have a taste of delicate and exquisite Japanese motif.

条形灯安装在过道的梯级上，方便客人上下，和侧壁的背投式玻璃灯的格调相统一。

散射的灯光运用在墙壁、包间、餐厅中，营造神秘而典雅的氛围，木质的桌椅、竹帘、榻榻米又增加了一种日式的纯朴与清新。

食客们徜徉于各个包间，宛如进入一个又一个迷幻的暗格，在品尝菜肴前就领略到日式风格的细腻与精巧。

Japan Dining

多明戈酒吧餐厅

Project > 项目名称 > 多明戈酒吧餐厅

Architect > 设计公司 > Tidy Arquitectos 设计公司（Albert Tidy、Ian Tidy ）

Location > 项目地点 > 美国

Built area > 项目面积 > 410 m^2

Photographer > 摄影师 > Sebastian Sepulveda

项目介绍 PROJECT DESCRIPTION

The project was awarded as first place finalists of a private architecture contest. The proposal consists in designing the interior of a space composed by a 60 m^2 access area, a double height volume of 150 m^2 located in a second level and an exterior terrace or patio of 200 m^2.

该作品曾在私人建筑比赛中获得一等奖，业主要求建造一个首层为 60 平方米的餐厅，二楼面积为 150 平方米，还包括 200 平方米的露台和天井。

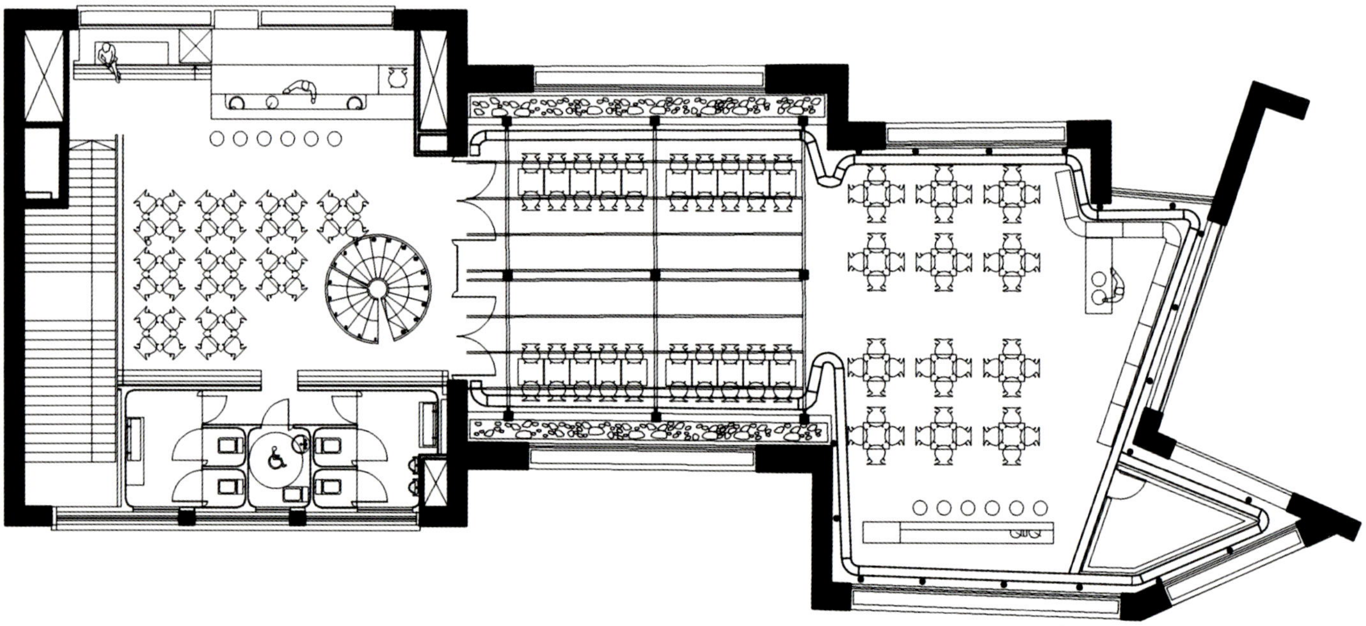

general floor plant level 2

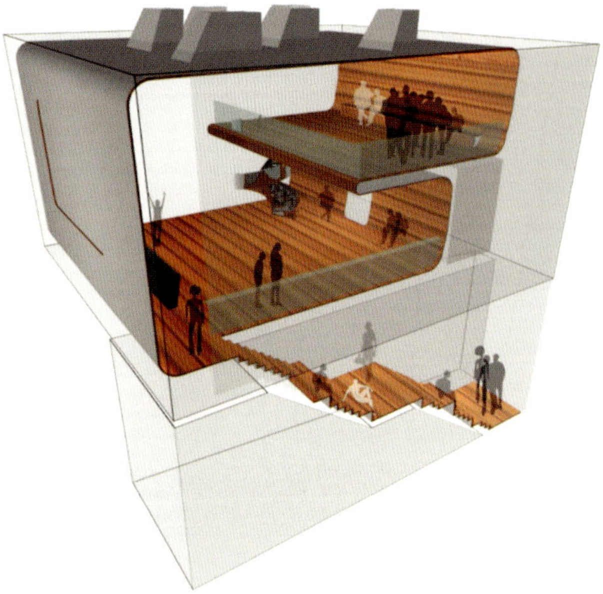

The enclosure is located in a well known and important commercial complex in the city of Santiago, Chile, in front of a high speed motorway. At night, images are projected from the interior onto a large translucid crystal window, marking its presence in the commercial complex.

餐厅坐落在智利圣地亚哥的一个知名的重要商业地段，正对一段高速公路。夜幕降临，餐厅内的投影仪将影像照射在一大块半透明的水晶窗户上，从外面看上去格外醒目。

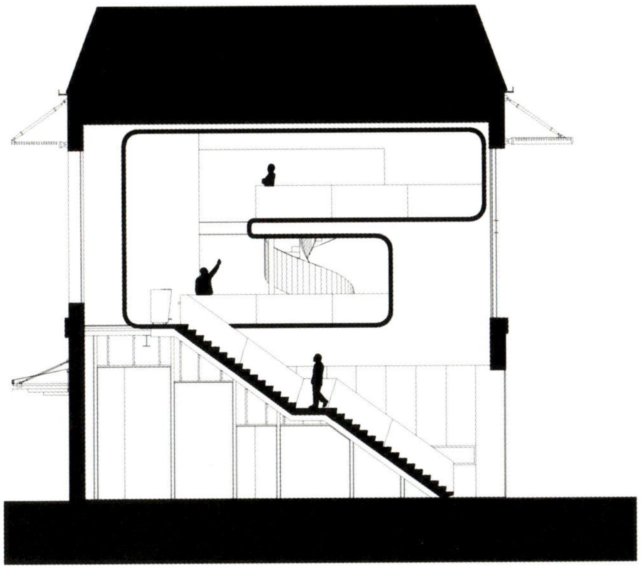

section A-A

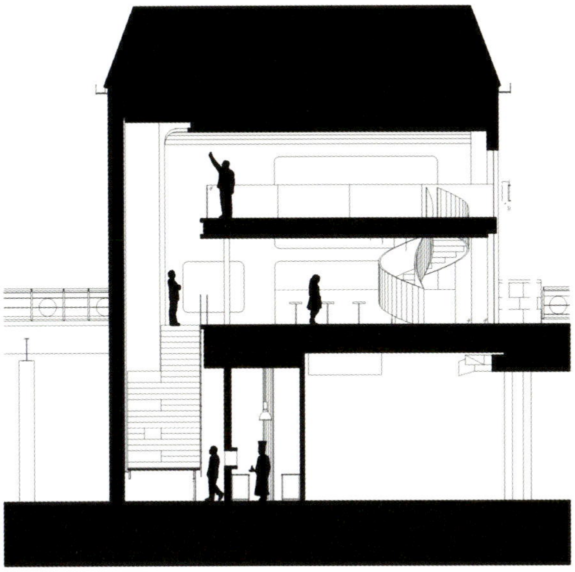

section B-B

Due to the visual noise caused by image saturation, natural to commercial areas, the project proposes a counterpoint which expresses itself as an abstract and isolated place from its context. This way, visitors are able to transport themselves to any scenario through the projection of digital images, such as: Tokyo, New York, Santiago; a Boeing's cabin or a cruise's bow.

为了配合光影视觉效果和周围的商业环境，设计师为餐厅打造了一个抽象且具有独立风格的用餐环境。数字投影可以变换场景，顾客仿佛置身于东京、纽约、圣地亚哥，甚至是波音客机的机舱或者是游轮的船首。

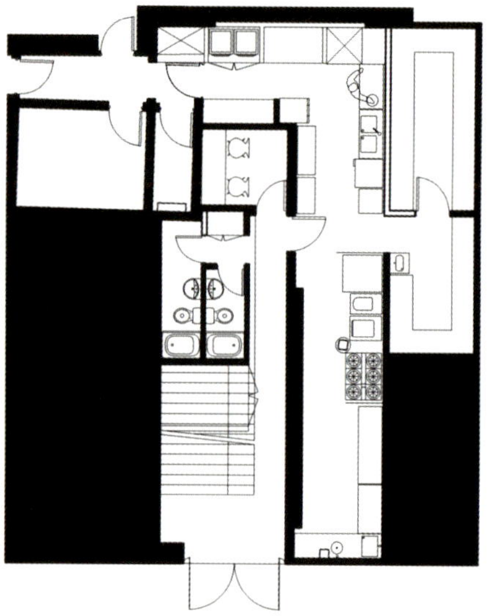

floor plant level 1

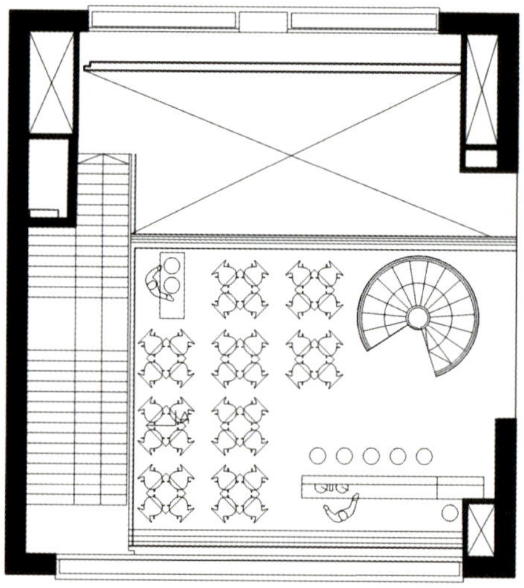

floor plant level 2

The main material chosen is ebony, which through folds, turns and gradations defines the place's space, emphasizing double heights, free floors and circulations, specific of buildings that base their activity on the act of seeing and being seen. This way, the access stairs turn into a "sloped plaza", the attic into a balcony and the kitchen into an inviting shop window.

空间的主材料运用了檀木，通过折叠、翻转，层次定位整体格调，两层加高的空间具有畅通的行径线。整个建筑的布局还注重观察与被观察的效果，楼梯的入口变成了一个“倾斜的广场”，阁楼成为楼厅，而厨房则作橱窗。

图书在版编目（CIP）数据

海外特色餐厅 / 深圳市海阅通文化传播有限公司 编 .
—南京 ： 江苏人民出版社， 2011.10
ISBN 978-7-214-07377-8

Ⅰ . ①海… Ⅱ . ①深… Ⅲ . ①餐厅－室内装饰设计－图集 Ⅳ . ① TU247.3-64

中国版本图书馆 CIP 数据核字（2011）第 183750 号

海外特色餐厅

深圳市海阅通文化传播有限公司 编

责任编辑：蒋卫国　张　蕊
责任监印：彭李君
版式设计：李　林
出　　版：江苏人民出版社（南京湖南路 1 号 A 楼　邮编：210009）
发　　行：天津凤凰空间文化传媒有限公司
销售电话：022-87893668
网　　址：http://www.ifengspace.cn
集团地址：凤凰出版传媒集团（南京湖南路 1 号 A 楼　邮编：210009）
经　　销：全国新华书店
印　　刷：深圳当纳利印刷有限公司
开　　本：965 mm×1270 mm 1/16
印　　张：20
字　　数：160 千字
版　　次：2011 年 10 月 第 1 版
印　　次：2011 年 10 月 第 1 次印刷
书　　号：ISBN 978-7-214-07377-8
定　　价：288.00 元　(USD 50.00)